GLOBAL SPECIFICATION AND VALIDATION OF EMBEDDED SYSTEMS

Global Specification and Validation of Embedded Systems

Integrating Heterogeneous Components

by

G. NICOLESCU
Ecole Polytechnique de Montreal,
QC, Canada

and

A.A JERRAYA
CEA-LETI,
Grenoble, France

Springer

A C.I.P. Catalogue record for this book is available from the Library of Congress.

ISBN 978-90-481-7550-5
ISBN 978-1-4020-6153-0 (e-book)

Published by Springer,
P.O. Box 17, 3300 AA Dordrecht, The Netherlands.

www.springer.com

Printed on acid-free paper

Contents

List of Figures

List of Tables

Chapter 1

Introduction

G. Nicolescu[1] and A.A. Jerraya[2]
[1]*Ecole Polytechnique de Montreal, QC, Canada,* [2]*CEA-LETI, Grenoble, France*

An embedded system is a system designed to perform a dedicated function, typically with tight real-time constraints, limited size, low cost, and low power. It is a combination of computer hardware and software as well as additional mechanical, optical, or other parts typically used in the role of actuators, sensors, and transducers. In some cases, embedded systems are part of a larger system or product, e.g. an antilock breaking system in a car. Examples of embedded systems are cell phones, digital cameras, GPS receivers, fax machines, printers, debit/credit card readers, heart monitors, blood gas monitors [Gan03].

Specification and validation are fundamental stages in the embedded system design process. The specification implies the definition of essential characteristics of a system and the role of the validation stage is to guarantee that the system complies with the requirements of the designers.

A key characteristic of modern embedded systems is the integration of heterogeneous components (hardware, software, mechanical, optical, etc.). This heterogeneity makes the global specification and validation of these systems extremely challenging, and these stages will require an increasing fraction of the overall design effort [ITRS06].

The purpose of this book is to provide a deep understanding of concepts and practices behind the global specification and validation of heterogeneous embedded systems. The book focuses on a key issue associated with these activities, the modeling of the interactions between heterogeneous components.

Modeling consists in providing a simplified representation of a system. The modeling approaches are as varied as the types of expertise required to design a system. Therefore, for the specialists defining new products, the model has to capture the functional and nonfunctional aspects of a system. For an architect, modeling helps analyze the system and partition it into several subsystems, and it

G. Nicolescu and A.A. Jerraya, Global Specification and Validation of Embedded Systems – Integrating Heterogeneous Components, 1–3.

assists in the choice of technologies for the different electronic, mechanical, or optical components. The design of the different parts of a system may require the modeling at different abstraction levels; the model captures only the information that is useful for a given design stage. Different models may be required for the different design phases (e.g. documentation, analysis, validation, refinement). However, common concepts (e.g. module, port, and interface) may be identified for the different models and application fields.

The different chapters of this book emphasize the importance and the challenges of modeling interconnections between heterogeneous components. A new solution needs to be devised to assist designers and simplify the current fastidious process. This solution implies a deep investigation and pragmatic definition of modeling basic concepts and the study of different types of heterogeneous interactions (hardware/software, continuous/discrete, and multiple abstractions). The results of this analysis then have to be synthesized in a generic approach providing heterogeneous interconnection models.

The second chapter of this book presents the common concepts of the different possible models for heterogeneous systems. It defines transversal basic concepts originating from different disciplines. The chapter also introduces the notion of a computing model. This model describes the functionality of a system as a hierarchical composition of basic computing elements.

The third chapter addresses the aspect of modeling for validation and presents the concepts specific to this stage. The notion of an execution model is introduced; this model describes the implementation of the functionality in a given technology (e.g. hardware or software). The execution models are the key concept in the validation design phase. The separation between the computing and execution models help highlight the problematic of modeling in the context of the composition of heterogeneous subsystems.

The fourth chapter presents the execution model–based validation technique for heterogeneous systems. The chapter starts by describing the anatomy and the main requirements of a global execution model for heterogeneous systems. The state of the art on heterogeneous systems validation is also presented.

The fifth chapter addresses a specific dimension of embedded systems heterogeneity, namely, hardware and software integration. It details the anatomy of a hardware/software execution model for heterogeneous systems validation. The chapter focuses on the hardware/software interaction which is generally provided by the operating system layer. After the introduction of the main techniques for embedded software validation and the state-of-the-art operating systems validation, an operating system execution model is proposed. The application of the operating system execution model for the design of a VDSL modem is presented.

The sixth chapter presents the anatomy of a continuous/discrete execution model for the timed execution of heterogeneous systems. The first part of the chapter presents the continuous and discrete execution models, and highlights the different adaptations required for their integration. After this study, the chapter focuses on continuous/discrete synchronization and the definition of a global execution model for embedded systems including both continuous and discrete subsystems. Finally, the chapter illustrates the global execution models and their role in the validation of two illustrative continuous/discrete embedded systems: a robot arm controller, a Σ/Δ converter, and a bottle filling system.

The seventh chapter defines a generic anatomy of execution models as well as a new methodology for the automatic generation of the execution models for heterogeneous embedded systems validation. The methodology is applied for the validation of three heterogeneous embedded systems: an optical switch, a VDSL modem, and an IS-95 CDMA system.

References

[Gan03] J. Gannsle, M. Barr, *Embedded System Dictionary*, CMC Books, 2003

[ITRS06] International Technology Roadmap for Semiconductor Design (ITRS), 2006

Chapter 2

Heterogeneous Systems Modeling: Basic Concepts

L. Kriaa
Tima Laboratory, Grenoble, France

1. Introduction

A model is a simplified representation of an existing, complex system. Modeling is the first step of any design and/or analysis activity related to the structure and behavior of a system. Modeling is not restricted to electronic systems. For example, nuclear fission simulations rely on atomic models while houses require architectural models.

Abstracting composition and refinement are fundamental issues in modeling activities. Any system "reasoning" or even any system manipulation requires modeling.

This chapter addresses the electronic systems whose implementations include hardware, software, and functional (unrefined) components.

Actual modeling approaches are as numerous as the necessary abilities required to design these systems. A model should permit designers of new devices to capture both functional and nonfunctional aspects of the system. For system architects, the modeling process means considering the partitions of the system and the choice of technology (e.g. software or hardware) that will permit the implementation of different partitions.

Designing the various subsystems of a complete system may require modeling at different levels of abstraction in order to hide more or less details depending on the design step under consideration. Different models may also be required for different design tasks (documentation, validation, analysis, and refinement). Nevertheless, all of these models and application domains share certain common elements.

These models are those that we tried to highlight through the use of transversal concepts, resulting from all of the disciplines and which, we hope, will be

G. Nicolescu and A.A. Jerraya (Eds.), Global Specification and Validation of Embedded Systems – Integrating Heterogeneous Components, 5–29.

understood by all. Thus, the drawback of our approach could pertain to the lack of rigor in the use of semantic concepts as certain words have different meanings depending on the context in which they are used. For example, the word “connection” has different referents if used by software designers or by those involved in the field of hardware. In addition, “communication” which evokes more or less complex processes, merges different elements according to simulations or synthesis targets.

Although modeling fields are numerous and interpretations vary, they obviously include common elements. Indeed, the objects used represent solutions to fundamental concepts.

To illustrate and avoid useless repetitions, in the first part of this chapter, basic modeling concepts are analyzed. The second section presents concepts such as system, module, interface, port, interconnection, and communication. This chapter highlights the need to distinguish between computation and execution models: the first allows the description of the system’s behavior in terms of hierarchical composition of basic computation elements, whereas the latter determines how computation is carried out using a particular technology. In this chapter, only computation models are described in detail as they are dedicated to system specifications. Execution models will be further detailed in the following chapter.

2. Basic Concepts for Specification of Heterogeneous Systems

This section is dedicated to the analysis of basic concepts used in heterogeneous systems on chip. This analysis can be considered an inventory of the current terminology used to define heterogeneous systems. First, basic elements of heterogeneous systems are defined. These basic definitions are then followed by a comparative study of two other closely linked concepts: models of computation and models of execution.

2.1 System

The term *system* refers to an entity composed of a set of interconnected hierarchical modules that communicate with their environment through well-defined interfaces [Lam02].

2.2 Module

A *module* is an entity that communicates through a well-defined interface with a fixed, external environment. A module is composed of contents and an interface. The module interface is connected to a communication network. A module may be elementary or hierarchical. An elementary module represents a leaf in

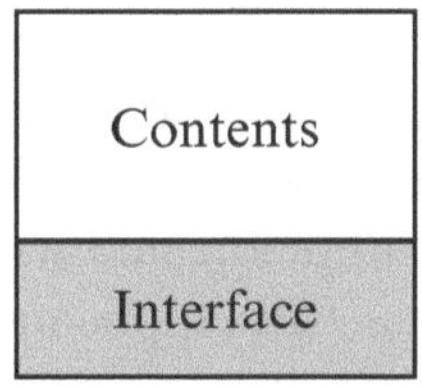

Figure 2.1. Module representation

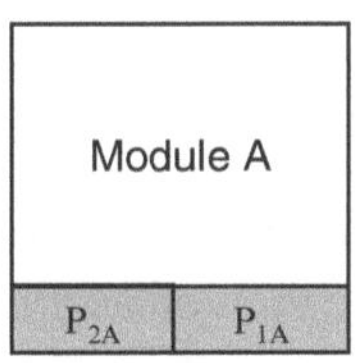

(a) Graphic representation of Module A

```
SC_MODULE (A) {
/*Interface*/

          sc_out<double>  P2A;
          sc_in<int>      P1A;

/*Behavior*/
          void behavior();
          SC_CTOR (A)
          {

          SC_THREAD (behavior);

          };
};
```

(b) SystemC description of Module A

Figure 2.2. SystemC description of Module A (behavior and interface)

the overall hierarchy and encapsulates an elementary behavior (task, process, function, etc.).

A *hierarchical module* consists of a set of elementary or hierarchical modules which are interconnected. Hierarchical modules can be used with other systems or within the same system. As such, a system may be considered a hierarchical module. A graphic representation of a module is shown in Figure 2.1.

Figure 2.2 shows a module made up of contents and an interface, using the SystemC language [SyC02]. Contents consists of the "behavior" process [denoted by SC_THREAD(behavior)]. The interface is explicitly represented by a set of ports (P1A, P2A) which are elementary ports (see Section 2.3). This example illustrates two port models: input ports are denoted by "sc_in" and "sc_out" indicates output ports.

2.3 Interface

An *interface* refers to a set of access points which allow modules to interact with both the environment and the remainder of the system.

The interface description may be more or less detailed. Depending on the level of abstraction, such details should include information about how components can be accessed and how such components may access the remainder of the system. The interface is seen differently throughout various abstraction levels. It may be described by some circuit pins, some other C function parameters or some procedures implemented within the modules. In the latter case, we are in the presence of an API (application programming interface).

An interface is defined as a set of ports and/or services depending on the manner in which components are accessed:

- *Services* define a high level of access generally performed by the APIs or the method calls of functions. They are used by high-level languages such as interface description language (IDL) [Cor04], C++ [Dup03], Java, etc. They are graphically represented by a rhombus (Figure 2.4a).

- *Physical ports* refer to a physical access such as an access to physical gates (e.g. "AND/OR" gates). Most HDLs (such as VHDL [Vhd93], SystemC [Hav02], etc.) can offer the interface description as physical ports. Graphically, they are represented by octagons (Figure 2.4c).

- *Logical ports* refer to the logical access to a module which holds a list of services (provided and/or required). They can be viewed as an abstraction of physical ports (e.g. VHDL ports handling integer data). Logical ports may be hierarchical. Services associated with logical ports contain the communication actions to allow module interactions with the external environment. Thus, a logical port is seen as $\{\mathrm{P_I A_{JI}}, \mathrm{J} = 0 \ldots \mathrm{M}, \mathrm{I} = 0 \ldots \mathrm{N}\}$ where $\mathrm{P_I}$ indicates port number I, and $\mathrm{A_{JI}}$ refers to a service provided by logical port I (e.g. a reading/writing service). An access point is defined in order to reach a logical port $\mathrm{P_I}$, noted port access point (PAP). In addition, a service access point (SAP) is defined in order to reach the services associated with the port (Figure 2.3). To illustrate, logical ports can only be depicted by squares (Figure 2.4b).

Figure 2.4 shows the types of interfaces defined above.

Figure 2.5 presents an example of an interface description as logical ports using VHDL language.

Module A can communicate with the external environment using two elementary logical ports (can be represented by pins of a circuit implementing A): $\mathrm{P_{1A}}$ (dealing with inputs) and $\mathrm{P_{2A}}$ (dealing with outputs). Services

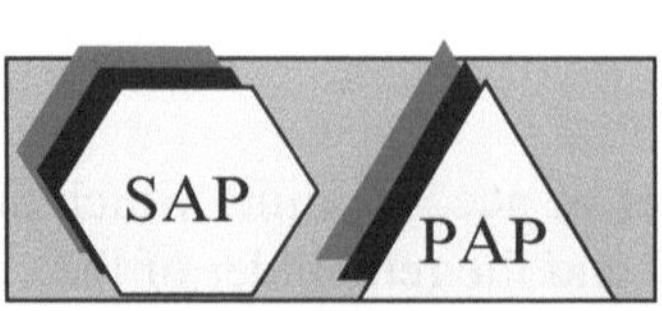

Figure 2.3. Logical Port

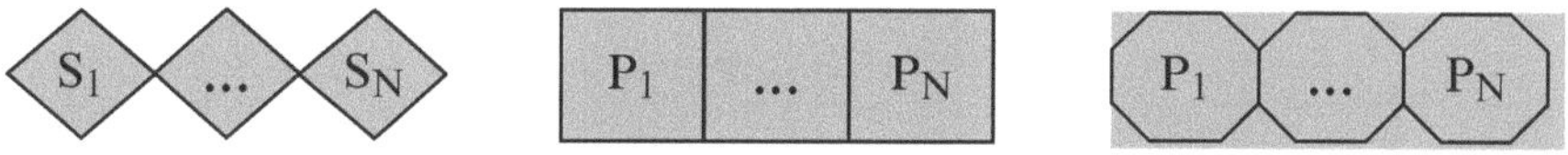

(a) (b) (c)

Figure 2.4. Interface as a set of: (a) services, (b) logical ports, (c) physical ports

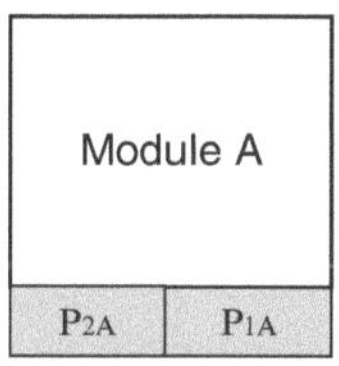

(a) Graphic representation of Module A

```
/*Module A Interface using VHDL */
ENTITY A is
port(
                P1A : in integer;
                P2A : out real;
    )
end A;
```

(b) VHDL description of Module A

Figure 2.5. Example of interface description

associated with logical ports are defined implicitly thanks to a library description of VHDL ports (i.e. stdlogic). The associated services are reading and writing.

Ports and services can be elementary or hierarchical (i.e. including several ports and/or services which share the same semantics, e.g. physical ports which share the same communication protocol). A system description requires a declaration of the modules which compose it and their interaction through their corresponding interfaces. Figure 2.6 shows a system composed of three modules called A, B, and C. The modules communicate through physical ports. Modules A and B communicate through their respective ports P1A and P1B using a specific communication protocol (e.g. full handshake protocol). This protocol models data flow transfer through three signals on separate wires. These signals are: (1) a signal used to establish transfers between both modules (i.e. Rdy), (2) a signal which transports data (i.e. Data), and (3) a signal used to send acknowledgment (i.e. Ack). This protocol model requires three physical ports. They can also be grouped into a single port. In that case, such a port would be known as a hierarchical port, such as the P_{1B} port shown in Figure 2.6.

Figure 2.7 presents different types of interfaces used to design a system.

To understand module interactions, it is necessary to clarify that "interconnection" indicates the mapping of two interfaces while "communication" refers to the manner in which the information is exchanged between modules.

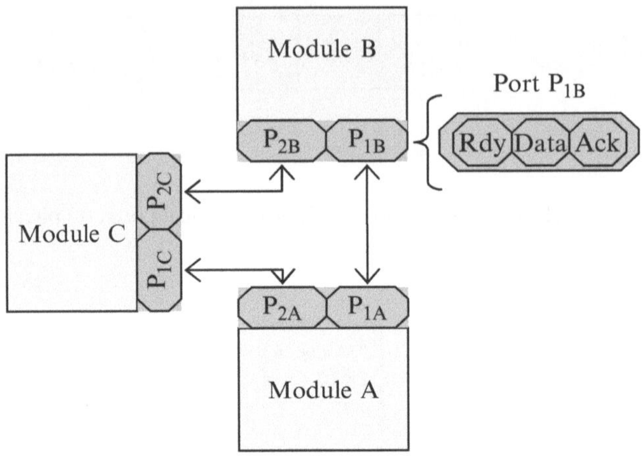

Figure 2.6. Illustration of a system including a hierarchical port

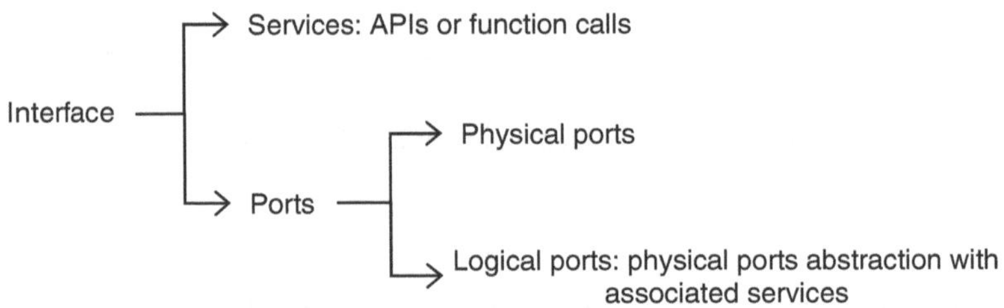

Figure 2.7. Interface definitions

2.4 Interconnection

Interconnection enables the interaction of several modules to interact while keeping their operations interdependent [Tsi03]. The interaction between the composed systems is defined as the associations (relationships) between various interfaces (ports and/or services) of different modules.

Interconnection is performed by means of connection which can be networks (e.g. Æthereal from Philips, Octagon from STMicroelectronics), abstract channels (e.g. FIFO channels, MPI channels, Bus ORB), or physical links (e.g. physical wires) according to the abstraction level. Interconnection ensures that data is transmitted between the modules. It can include transparent communication details for the modules it interconnects. Interconnection is also defined as a "communication channel" and is defined according to:

- The type of media which conveys information (e.g. physical wire or networks);
- The data type transmitted through the interconnection (e.g. generic data or fixed-point data);

- The channel behavior (procedure/manner used to convey data through the media). Thus, an abstract interconnection (abstract communication channel) can be refined by a particular module. However, separation between both module and channel can be of interest to the design flow. In fact, the refinement of communication and computation processes generally requires different techniques.

The example shown in Figure 2.9 specifies a SystemC description for a three module system connected through point-to-point interconnections (Figure 2.8). Each module has two logical ports. Module A has an interface composed of the input P1A port (that conveys integer data types) and the output P2A port (that conveys double data types). The Module C interface is composed of two logical ports: P1C and P2C input and output ports respectively that transfer double data types. The interface of Module B is composed of the output P_{1B} output port transferring double data type and the P_{2B} input port conveying a double data type. Interconnection occurs through logical wires represented by "sc_signal" in SystemC language. The associations between ports occur as follows: P_{1A} with P_{1B}, P_{2A} with P_{1C}, and P_{2B} with P_{2C}.

Using the concept of association, a new concept, namely compatibility of interfaces, has been defined. Two interfaces, e.g. I1 and I2, are compatible if they can be linked by a simple connection. It refers to a generalization of the connection rules used in hardware description languages. For example, input and output ports which share the same structure can be connected through a simple physical or logical connection whose role is entirely dedicated to conveying information between both ports. To extend this concept to interfaces such as those defined above, it is necessary to expand this concept to include logical ports, services, and hierarchical ports. Thus, the simple association of two interfaces, in general, requires that their different elements share the same

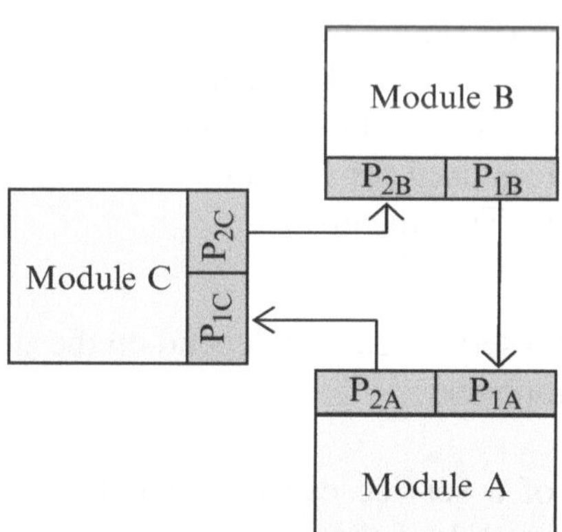

Figure 2.8. Interconnections amongst three modules using logical port associations

```
sc_main (argc, argv)
{
/* signal declaration*/
sc_signal <double> S1, S2;
sc_signal <int> S3;

/* Module Instantiation */

A M_A("module A");
B M_B("module B");
C M_C("module C");

/* A and B interconnection through P1A et P1B*/
M_A.P1A(S3);
M_B.P1B(S3);

/* A and C interconnection through P2A et P1C*/

M_A.P2A(S2);
M_C.P1C(S2);

/* B and C interconnection through P2B et P2C*/

M_B.P2B(S1);
M_C.P2C(S1);

/*remaining behavior*/}
```

Figure 2.9. Example of SystemC interconnections

hierarchy and nature (service, logic, or physical), i.e. all the ports/services are compatible as pairs. If ports (logical or physical) and services are hierarchical, the interconnection is considered a hierarchical link.

To extend compatibility definitions to services, the following assumption is made: the defined services used for communication are independent from the module which uses them. That means:

- The execution of a service cannot depend on the state of the module, but simply on its parameters;
- The refinement of a service can be carried out without changing the description of a module if the refinement respects the API used by the module.

There are two types of interconnection: homogeneous and heterogeneous.

Homogeneous interconnections. Homogeneous intercommunions allow the association of compatible interfaces. There are three different categories of homogeneous interconnections:

- *Physical or explicit interconnections* are associated with physical ports. They can be defined as connection rules defined in hardware description languages (Figures 2.16 and 2.17).
- *Logical interconnections* refer to associations of logical ports. They present associations between the services and ports of the interfaces concerned. They can be viewed as the physical grouping of interconnections in a more abstract interconnection or a refinement of an abstract interconnection (Figure 2.15).
- *Abstract or implicit interconnections* are defined as the association of services from various communication interfaces. For example, a system can be seen as an assembly of modules communicating through API (function call) without specifying the nature of the interconnection. This interconnection is generally hidden using an abstract model (see Figure 2.10). For example, a system described using the parallel description language named MPI (Message Passing Interface) [Gro96]. This system is composed of two modules: a transmitter and a receiver. The interconnection between these two modules is hidden behind the API communication (in the example shown in Figure 2.11, communication services are defined by "MPI_send" and "MPI_Recv" API). In this case, this link is considered to be a logical interconnection with an abstract behavior and a compatible data type.

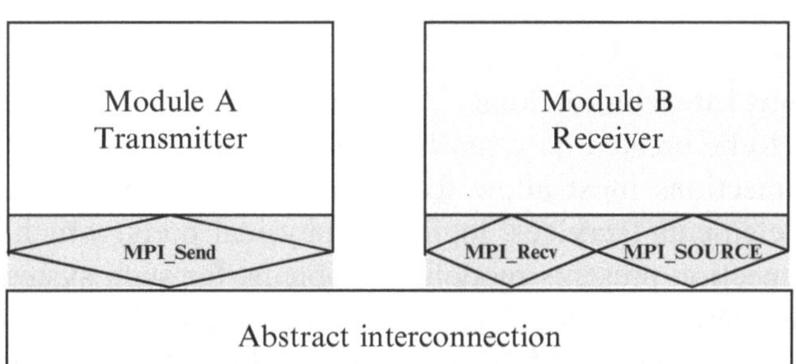

Figure 2.10. Abstract interconnection

```
main(int argc, char *argv[])
{
      /* parameters declaration for the program */
      const int tag;
int id, ntasks, source_id, dest_id, err, i;
     MPI_Status status;
     int msg[2];
  /* MPI instructions: initialization, …*/
/* testing the number of system modules */
if (ntasks != 2) {
    printf("You must use two processors to run this program\n");
    MPI_Finalize();                      /* Quit if there is only one
processor */
    exit(0);
  }
  /* Module B: the receiver */

  if (id == 0 ||id ==2) {
    for (i=1; i<ntasks; i++)
          {
      err = MPI_Recv(msg, 2, MPI_INT, MPI_ANY_SOURCE, tag,
            MPI_COMM_WORLD, &status);         /* receiving the message
*/
      source_id= status.MPI_SOURCE;             /* getting the
transmitter id */
      printf("Received message %d %d from process %d\n", msg[0], msg[1],
source_id);
    }
}
  /* Module A: the transmitter  */

else if (id ==1)
{
    msg[0] = id;              /* message identifier */
    msg[1] = ntasks;          /*identify the number of processus*/
    dest_id = 0;              /* destination address */
    err = MPI_Send(msg, 2, MPI_INT, dest_id, tag, MPI_COMM_WORLD);
  }

  err = MPI_Finalize();          /* end MPI */
  if (id==0) printf("Ready\n");
  exit(0); }
```

Figure 2.11. MPI code describing an abstract interconnection

Heterogeneous interconnections. Heterogeneous intercommunions allow relationships to be created between incompatible interfaces. The behavior of such interconnections must allow for the exchange of information between the different elements (services, logical or physical ports) which it connects. This interconnection presents modeling problems for such systems and their execution (simulation). Figure 2.12 shows an example of a system composed of two modules, A and B, which have incompatible interfaces. Module A has three logical ports (P_{1A}, P_{2A}, and P_{3A} are logical ports which convey "integer" type data) whereas Module B has two physical ports (P_{1B} and P_{2B} convey "bit"

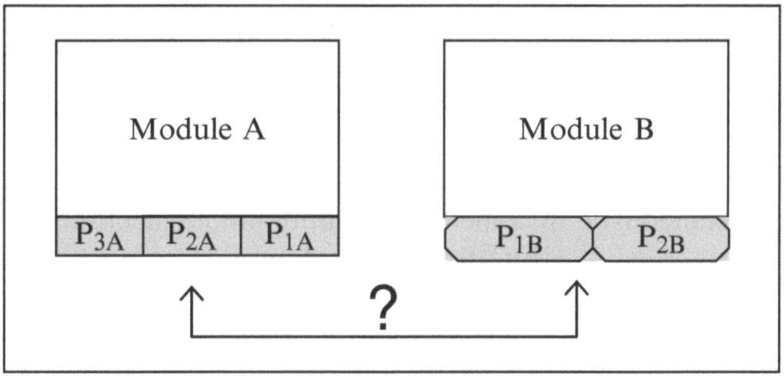

Figure 2.12. A system with heterogeneous interconnection

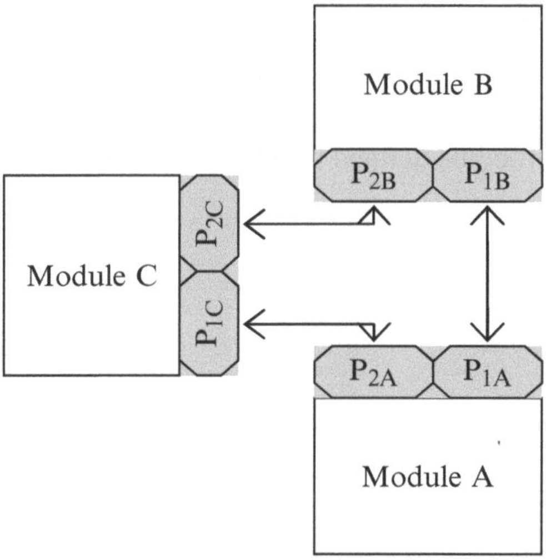

Figure 2.13. Point-to-point interconnection

type data). This system thus requires heterogeneous interconnection modeling to link both modules.

As defined above, interconnection between the different modules can be carried out according to various topologies: point-to-point or multipoint. In all cases, this interconnection will be represented by a net or a network of bound wires; it acts as a generalization of the representation of the "Net" in hardware description languages (HDL).

Point-to-point interconnection. A point-to-point interconnection represents a direct link between exactly two interfaces. For example, the link will be connected through a signal between the two concerned physical ports concerned. Figure 2.13 illustrates point-to-point interconnection (e.g. the elementary physical ports P_{1B}, P_{2B} of Module B are linked directly with the elementary physical ports P_{1A} of Module A and port P_{2C} of Module C respectively).

Multipoint interconnection. Multipoint interconnections associate several interfaces simultaneously. In some cases, multipoint connections can generate conflicts and thus require arbitration. According to the abstraction level, the arbitration function can be as simple as the choice of a single link among n or more complex and thus requiring routing function implementations or the negotiation of services. Figure 2.14 illustrates an example of a simple multipoint interconnection. The SystemC description of this system is given in Figure 2.15.

In this example, the interconnection is presented as a FIFO channel described at a high level of abstraction. The module interfaces are defined as elementary logical ports I_A, I_B, and I_C for Modules A, B, and C respectively.

In the remainder of this chapter, the homogeneous interconnections are considered to be complex. Nevertheless, they can hide complex behaviors and can be represented by a module defined by a behavior and an interface. Figure 2.16

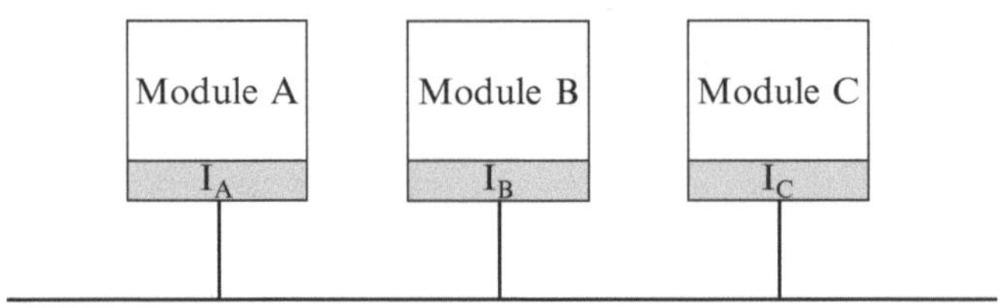

Figure 2.14. An example of a simple multipoint interconnection

```
sc_main(argc, argv)
{
/* Declaration of a FIFO interconnection */
sc_fifo Fifo;

/* Module instantiation */

A M_A("module A");
B M_B("module B");
C M_C("module C");

/* Interfaces associations */
M_A.IA(Fifo);
M_B.IB(Fifo);
M_C.IC(Fifo);

/*reste du comportement*/}
```

Figure 2.15. SystemC description of a simple multipoint interconnection

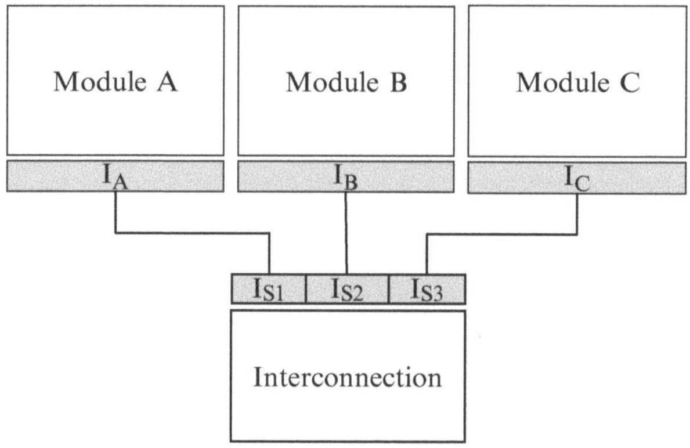

Figure 2.16. Complex multipoint interconnections

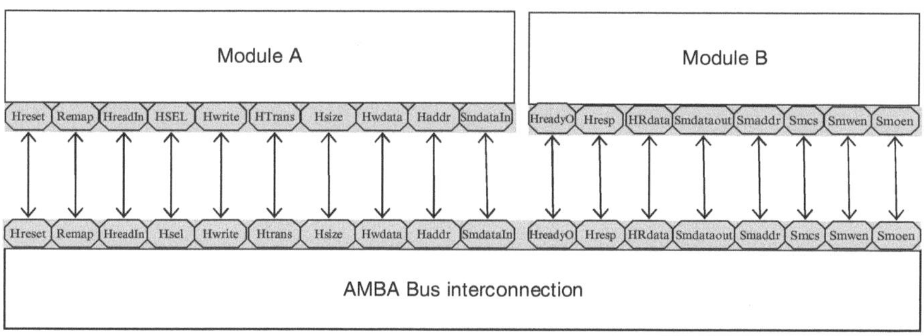

Figure 2.17. AMBA bus interconnection

illustrates a multipoint interconnection where the link is considered as a particular module. The interface of a specific interconnection module (IS1, IS2, IS3) associates all of the ports for Modules A, B, and C, and deals with the routing required for transferring information amongst the various modules.

The use of a bus interconnection is considered as a complex multipoint interconnection. Figure 2.17 represents the interconnections of two modules, A and B, through an AMBA Bus indicated by AHB [Ahb99]. The SystemC code for this illustration is shown in Figure 2.18. Modules A and B communicate through their physical ports (ten ports for Module A and eight ports for Module B).

2.5 Communication

Communication consists in the exchange, or the transfer, of information amongst the various modules of a system while respecting preestablished conventions.

```
// Signal connexion declaration
  sc_signal<bool> RESETn, RMP, READYIn, SEL, WRITE, READYOut;
  sc_signal<sc_uint<2> > TRANS, RESP;;
  sc_signal<sc_uint<3> > SIZE;
  sc_signal<sc_uint<32> > WDATA, ADDR, DATAIn, RDATA, DATAOut;
  sc_signal<sc_uint<26> > SADDR;
  sc_signal<sc_uint<8> > CS;
  sc_signal<bool> WEn, OEn;
// Module instantiation
  A M_A("Module A");
  B M_B("Module B");
// Interconnection module declaration
  AHB smi("AHBSmi");
// Define relationships amongst modules and interconnection
  M_A.HRESETn(RESETn);
  M_A.REMAP(RMP);
  M_A.HREADYIn(READYIn);
  M_A.HSEL(SEL);
  M_A.HWRITE(WRITE);
  M_A.HTRANS(TRANS);
  M_A.HSIZE(SIZE);
  M_A.HWDATA(WDATA);
  M_A.HADDR(ADDR);
  M_A.SMDATAIn(DATAIn);

  M_B.HREADYOut(READYOut);
  M_B.HRESP(RESP);
  M_B.HRDATA(RDATA);
  M_B.SMDATAOut(DATAOut);
  M_B.SMADDR(SADDR);
  M_B.SMCS(CS);
  M_B.SMWEn(WEn);
  M_B.SMOEn(OEn);
// Connection of Module AHB with the remainder of the modules
  // With Module A
  smi.HRESETn(RESETn);
  smi.REMAP(RMP);
  smi.HREADYIn(READYIn);
  smi.HSEL(SEL);
  smi.HWRITE(WRITE);
  smi.HTRANS(TRANS);
  smi.HSIZE(SIZE);
  smi.HWDATA(WDATA);
  smi.HADDR(ADDR);
  // With Module B
  smi.SMDATAIn(DATAIn);
  smi.HREADYOut(READYOut);
  smi.HRESP(RESP);
  smi.HRDATA(RDATA);
  smi.SMDATAOut(DATAOut);
  smi.SMADDR(SADDR);
  smi.SMCS(CS);
  smi.SMWEn(WEn);
  smi.SMOEn(OEn);
```

Figure 2.18. SystemC description of a system with AMBA bus interconnections

We will assume that communication occurs through port-associated actions. Thus, for a given module, communicative actions within the ports gives the API access to the remainder of the system and the external environment. The definition of this API constitutes an architectural decision. It enables the separation between the behavior of the module and the services provided by the module environment. This is commonly referred to as separation of concerns or communication and computation orthogonality [Lee99].

2.6 Hierarchy

The notion of hierarchy in computer science data is often associated with the concept of inheritance (expressed primarily in relational models). In electronic systems, this definition is closely linked to that of "composition".

2.7 Composition

Composition means an organized collection of interacting components designed to achieve a coherent and common behavior [Tsi03].

A composed system consists of a set of interconnected modules. Each module can be an elementary module or it can be composed of a set of modules (called "submodules").

Thus, a composed system is defined as a composition where interconnections are used to associate the interfaces of its different modules. Figures 2.19 and 2.20 illustrate an example of Module C, which is composed of two modules, labeled D and F. The interface of Module C is defined as two logical ports PC1 and PC2. The interfaces of Module D and F are composed of logical ports PD and PF respectively. Ports PC1 and PC2 are connected to ports PD and PF respectively.

Figure 2.21 provides a clear example which illustrates these basic concepts. The complete system is represented by a module which interacts via its interface modeled by logical elementary ports P_{1S} and P_{2S}. This system is composed of

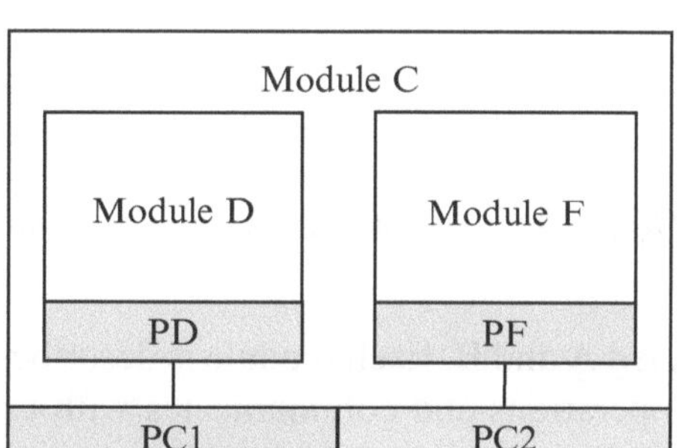

Figure 2.19. A graphic representation of a composed system

```
SC_MODULE (C)
{
// Module C ports
sc_in<int> PC1;
sc_out<int> PC2;


/* Sub modules declaration*/
D *M_D;
F *M_F;

/* Module C constructor*/
SC_CTOR ( C )
{
/* Sub modules instantiation*/

M_D= new D ("Module D");
M_F= new F ("Module F");

/* Associations between ports PD, PF and Module C ports*/

M_D->PD(PC1);
M_F->PF(PC2);
};
}
```

Figure 2.20. SystemC description of a composed system

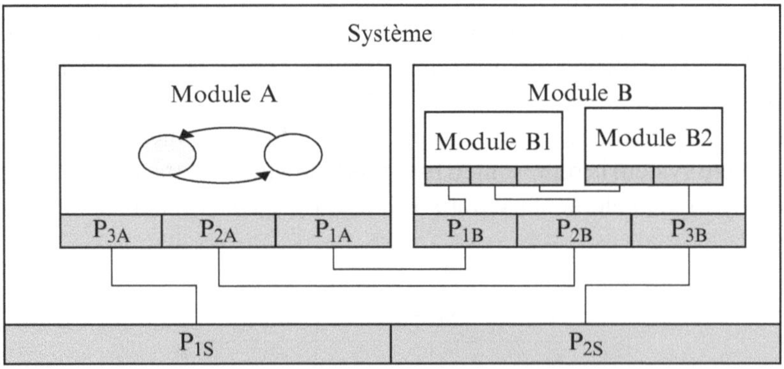

Figure 2.21. Global specifications of system

two submodules, labeled A and B. Each module is described by a behavior representing arithmetic operations and communicates with external environment through a specific interface. Module A is an elementary module; its behavior is described by two tasks. Module B is hierarchical and contains two submodules (B1 and B2). The different system modules communicate using the logical ports linked through logical interconnections (in this figure, Modules A and B communicate via two interconnections between ports P_{1A} and P_{1B} and ports P_{2A} and P_{2B}).

3. Computation Model vs. Execution Model

One of the main challenges in the study of embedded systems consists of the ability to separate what is done by the system from the way it is done, referring to the implementation of the system. These two concepts will be called "computation model" and "execution model". This section will address these two modeling concepts.

Computation models are primarily used to provide a description or a representation of the systems; for example, a computation model of a filter can be described mathematically, using differential equations. However, the *execution model* of the he same filter represents an implementation of its computation model. It can refer to pure hardware implementations using registers, or pure software using a processor or simply a description of the mathematical function using a programming language, called, in such a case, a functional execution model.

Several studies have defined these two concepts. For example, in the study carried out by [Sim98], the computation model involves a high level of abstraction. It is defined by three concepts: (1) the basic objects (data, operations, etc.); (2) the composition semantics of the operations (procedural or declarative); and (3) the execution order (directed by the data or control). This definition allows to model implicitly the manner in which the system carries out these functions [Sim98].

Other studies, such as those conducted by [Lee99], [Liu01], and [Sgr00], focus on the formal definitions of the different computation models. Components (modules) and interactions are determined for each model. Interactions represent the way communication can be established between the modules (asynchronous, synchronous, in continuous time, discrete events, etc.). These elements are presented as mathematical concepts in order to define a simulation model for each computational model.

All of these definitions have common semantics with the behavior execution. They do not carry the final implementation of the system (through the execution models). This gap between the computation model and the execution model constitutes a source of error for the final system implementation as well as a limitation for the exploration of the architecture.

It is thus necessary to introduce separate definitions for computation and execution models in order to reduce the gap between implementation and modeling. The key issue is to study separately the computation (behavior) and the interconnection, as well as the different abstraction levels of the computation models and the execution models.

The following sections define the computation models and present the computation models for the components and the interactions of a system. The composition of computation models is also explained.

4. Computation Model

The computation model presents the way in which a result is obtained independently of the time and data handled. This is considered a description of the behavior [Jan03] and makes it possible to determine the relationships between input and output values.

4.1 Elementary Computation

An elementary computation is defined as a set of data transformation rules using mathematical operations. This elementary calculation is defined by:

- Inputs
- Outputs
- Transformation functions, which can be combinatory functions (without states) or sequential functions (with states). Such functions can be simple or complex depending on the granularity of computations.

A computation model is defined as a set of rules for the composition of elementary computations (at different levels of granularity) in order to carry out the set of transformations on the objects (data). According to this definition, computation models allow to describe system operations by specifying the objects' sets, the operations performed on such objects as well as the order in which they occur. Objects are thus defined by sets of inputs, outputs and the operations on these objects (transformation functions). Relationships between these objects can be explicit (specified explicitly by users through operations) or implicit (hidden by the transformation functions).

Basically, the description of these systems is made through primary computation models. These models are classified according to (1) the order of occurrence of the elementary computation and (2) the interference between the orders of various computations. There are two different categories:

- *Computation model directed by data*: in this model, the order in which the operations occur depends on the availability of the data. It is generally represented by a data flow graph (DFG) or differential equations.
- *Computation model directed by control*: the order in which the operations occur in these models depends mainly on the current state of the system. It is generally represented by control flow graphs or finite state machines.

Moreover, there is a mixed representation of both models (control and data). They are represented by graphs known as control data flow graphs (CDFG) [CDF05].

However, these models become inadequate when representing composed systems. This is due to the complexity of these systems and the simultaneous use of different models. By abstracting computation in an elementary module and presenting only the composition of the system, hierarchical models, and more particularly distributed models, are introduced in order to describe complex systems. This model is presented in detail in Section 4.5.

4.2 Basic Computation Model

This section describes the most popular types of basic computation models. Each model defines specific characteristics pertaining to the set of notations employed to express the operations. The execution order of these operations, such as concurrency and parallelism, are also addressed.

4.3 Data Flow Model

A data flow (DF) model is a data-driven model. It represents computations where the operation sequence is defined according to the availability of the data [Ger93; Naj99]. Several DF models have been defined. The most common ones are DFGs and differential equations. The main characteristic of DF models lies in the absence of a global state of the system. A DF model then becomes the equivalent of a combinatory circuit [Ger93; Mad97].

Data flow graph. The data flow graph (DFG) is a generic model that allows the representation of the data dependencies. It is composed of a set of nodes and arcs: nodes correspond to data-controlled operations; arcs represent data values. Thus, the objects defined by this model are the sets of nodes acting on the input to provide output. The relationships between these objects are implicitly described by sets of arcs which connect nodes. This model includes several manners for activating the nodes [Jag96]. The various types are differentiated by activation rules of the different nodes. There are two models:

- Data-driven DF model: the nodes are activated iff all of the input data are available.
- Demand-driven DF model: a node is activated only if the node is required to complete a given computation.

Using this graphic presentation, data dependency is clarified and the determination of parallelism and the order in which operations occur are commonplace.

DFG models are of interest as they are based on formal models which allow for complex mathematical reasoning. Thus, formal transformations of DFGs are easily defined and can be employed to model several systems such as the signal processing systems.

Differential equations. Differential equations are mathematical notations which express computations such as the DFG in the form of equations which involve an unknown function and its derivatives. The operations consist of the functions and their equality operations. The order of execution is implicitly defined by the data dependence.

4.4 Control Flow Model

In the control flow model, a precedence relation defines the order in which operations are executed. The execution of a given operation is conditioned by the execution of a previous operation. The models which are most often used to represent the control flow models are the finite state machines and the control flow graph.

The main characteristic of this model relates to the concept of system variables and global state. In fact, for any computation time, the control model is characterized by its current state as well as the state of the data which it handles [Hol02].

Finite state machine model. The finite sate machine model (FSM) is a directed graph whose nodes represent the states of the system and whose arcs represent transitions [Gir99]. Several models of FSMs can be defined according to the nature of the states and transitions.

A state can contain complex computations, even internal states. Conditions are associated with transitions. Moreover, aside from conditions, transitions can also carry out complex computations. The FSM includes an initial state. It presents the starting point of computations. State changes depend on the transition functions associated with the different arcs and the state where the last operation was carried out. FSM models are generally used to model the control parts of the system.

Control flow graph. The control flow graph (CFG) model describes a computation sequence of the operations with a directed graph which shows a set of nodes and arcs. The nodes in this model represent the operations; the arcs indicate the precedence relations between these operations and can thus be used to define the computation sequence. The order in which nodes occur depends primarily on the computation of the previous node. Certain conditions can be associated with the arcs to validate the occurrence of the following operations.

4.5 Computation Model Composition

To handle complexity, systems are generally modeled using the composition of modules. Each module can be described as an independent entity that uses specific computation model (as previously defined). The current trend is the design modules independently from their implementation environment. This technique enables the reuse of the same modules in different contexts or with different platforms.

Hierarchical models make it possible to describe composed systems. There are two different models:

- *The Behavioral (Specification) Hierarchy* simply facilitates syntax. Objects from this model can hide basic computation models such as those described above. An illustrative example of this type of model is the hierarchical FSM [Ard99; Lut00; Mar91] called the extended FSM and hierarchical DF models [Mar91]. It is even possible to imagine combining the DF and the control flow models to describe a system. In this case, the model would be considered to be a hybrid. The main restriction of such hybrid models resides in their difficulty to describe efficiently distributed computations between several modules. Thus, if a DF model is used to specify control flow objects, the reasoning in terms of mathematical operations will be difficult due to possible loops in the control nodes. If a control flow model is used to describe DF objects, the difficulty will consist of expressing parallelisms between various DF modules' objects. In the literature, all authors of experiments conducted in this manner end up reducing the model of the entire system to a basic finite state machine in order to analyze it.

- *The Interconnection Hierarchy* defines the composition of a system as interconnections between the various modules used for its creation. It consists of abstracting the module behaviors and specifying only the interaction sets of each module. The remainder of this section will solely address these models.

Distributed models are defined as a composition of several interconnected modules (Figure 2.21). They comprise two types of objects: modules and interconnections. Each object is associated with connection terminology as defined in Section 2.4. Basic composition relations are associated with object interfaces. In fact, in a distributed model, each module is an independent entity and communicates with other components only through its interface. Interconnections amongst modules allow for computational coordination of different modules. Coordination can also be formulated using a computation model.

Computation models of distributed systems are thus defined by the composition of two types of models:

- The components computation model (Section 4);
- The interconnection computation model (Section 5).

Composition is generally defined as the product of the state of the machines which represent the modules and their interconnections. This definition generates an exploded state number and makes rather difficult any formal reasoning by complex distributed systems. Computation models thus seem inadequate to carry out the various analyses required as well as the architectural exploration. In order to respond to current needs, execution models can be considered to be acceptable solutions.

5. Computation Model of Interconnection

Interconnection is defined as the association of various module interfaces (Section 2.4). Interconnection functionalities consist of carrying out information transfer amongst modules, such transfer being either simple (e.g. read/write) or complex (e.g. by means of a complex protocol such as "hand-shaking" or TCP).

To coordinate parallel modules, three types of functions have to be defined at the interconnection level. These three functions are: data transfer, synchronization, and quality of service (QoS) to ensure a well-defined quality of service. These three functions are described hereafter.

5.1 Data Transfer

Data transfer refers to the routing of data from one entity to another. This can occur in two different manners: shared memory or message passing.

Shared memory. The concept of shared memory allows for interactions amongst several modules which can reach the same memory segment. In general, it allows for rapid data exchange amongst processes. Data duplication becomes unnecessary as they are simply located in a place which is accessible by the different processes. Moreover, memory can be reached by several processes. However, this model can require complex synchronization to ensure data coherence.

Message passing. Message passing is defined as the transfer of messages from one module to another. In this case, each module is equipped with a local memory capacity. Such transfers are slower than those conducted with shared memory. However, synchronization mechanisms are simple and memory synchronization problems are avoided. Unfortunately, this model generates data duplications which contribute to increasing system latency.

5.2 Synchronization

Synchronization enables the coordination between various communicating modules in order to avoid conflicts which cause data loss. Synchronization can be conducted by a global agent (resource) such as the use of a clock or local resources such as semaphores and it can be either synchronous or asynchronous. Note that synchronization models are specific to a single interconnection. A system including many interconnections can thus contain several different synchronization models.

Synchronous model. The term "synchronous" comes from the combination of two Latin roots: "syn" which means "with" and "chronous" which refers to "time". It describes objects and events which are coordinated according to a temporal X-coordinate. A synchronous model supposes the existence of a central agent which synchronizes communicating modules. This model is generally characterized by a blocking mode, i.e. communication imposes dependent activities and wait time between the various modules concerned [Syn05].

Asynchronous model. The term "asynchronous" is derived from the Latin roots "asyn" meaning "without" and of "chronous" which refers to "time". It describes objects and events which are not coordinated in time. Asynchronous models are characterized by communications which are not influenced by the execution of the behavior of certain modules. For example, a module can continue its executions after sending a message (without having to wait for the receiver's acknowledgment). This nonblocking mode supports multitasking [Syn05].

5.3 Quality of Service

Quality of service (QoS) is a measure that enables interconnection performance analyses. Furthermore, it also directs transfer operations in order to guarantee that constraint sets pertain to the suggested services. The interconnection of quality of service is ensured by a set of control mechanisms which can be, for example, routing techniques and scheduling of the various data transmitted. Such mechanisms can be used in static or dynamic manners [QoS99].

6. Computation Model for Heterogeneous Systems

To define heterogeneous system computation models, components, and interconnected computation models must first be determined. Up to now, a single model which could provide a heterogeneous system description has yet to be defined. This is due to the use of several computation models (elementary, hierarchical, etc.). Thus, the definition of heterogeneous system computation models requires an evaluation of all system operations. This being hard to define (due to different concepts and various languages), the choice of a single model for these systems does not represent an optimal solution. To model such systems, the distributed model (defined in Section 4.5) seems to be an efficient solution. This choice comes from the definition of heterogeneous systems as a composition of various interconnected modules. Computation models are thus sets of different modules with different computation models (Section 4.2) and heterogeneous interconnections. However, the execution semantics of such a model can only be defined through the execution models of the different elements of a heterogeneous system (components and interconnections).

The Chapter 3 will focus on defining systems execution models through the components and interconnection execution model studies.

7. Conclusion

This chapter presented common terminology found in various studies which have addressed different types of models and specific languages. This terminology covers a certain number of concepts pertaining to these languages such as behavior, interface, interconnection, communication, hierarchy, and composition. The second part, dedicated to the importance of separating the concepts of computation and execution models, defines heterogeneous system models. Furthermore, these sections listed the components and interconnections of various computation models. Finally, difficulties inherent to heterogeneous system modeling were highlighted before the presentation of solutions.

References

[Ahb99] *AMBA Specification*, ARM Limited Edition, 1999, online http://www.gaisler.com/doc/amba.pdf.

[Ard99] L. Arditi, A. Bouali, H. Boufaied, G. Clave, L. Leblanc, R. De Dimone, "Using Esterel and Formal Methods to Increase the Confidence in the Functional Validation of a Commercial DSP", In Proceedings of ERCIM Workshop on Formal Methods for Industrial Critical Systems, Toronto, Italy, 1999.

[CDF05] Service online, "Modeling Mixed Control/Data Flow Systems Using PCC Models", http://www.iss.rwthaachen.de/Projekte/Tools/MOLIERE/pcc_intro/pcc_intro.html.

[Cor04] Service online, "Common Object Request Broker Architecture: Core Specification", v3.03.3, March 2004, OMG, Inc. http://www.omg.org/ technology/documents/corba_spec_catalog.htm.

[Dup03] S. Dupin, *Le language C++*, edited by Campus Press Edition, 2003.

[Ger93] G. Gerrit de Jong, *Generalized Data Flow Graphs Theory and Applications*, Technische Universiteit, Eindhoven, 1993.

[Gir99] A. Girlaut, B. Lee, E. Lee, *Finite State Machines with Multiple Concurrency Models*, Fellow, IEEE, 1999.

[Gro96] W. Gropp, E. Lusk, A. Skjellum, "Using MPI: Portable Parallel Programming with the Message-Passing Interface", MIT Press, 1996.

[Hav02] A. Haverinen, M. Leclercq, N, Weyrich, D. Wingard, "SystemC^{TM} based SOC Communication Modeling for the OCP^{TM} Protocol", White paper, vol. 1.0, October 2002.

[Hol02] G. Holloway, M.D. Smith, "The Machine-SUIF Control Flow Graph Library", Version 2.02.07.15, July 15, 2002.

[Jag96] R. Jagannathan, *Parallel and Distributed Computing Handbook*, Chap. 8: "Dataflow Models", pp 223–237, edited by Y. Zomaya, 1996.

[Jan03] A. Jantsch, Morgan Kaufmann, *Modeling Embedded Systems and SoC's: Concurrency and Time in Models of Computation*, Hard-Cover, June 2003.

[Lam02] L. Lamport, *Specifying Systems: The TLA+ Language and Tools for Hardware and Software Engineers*, Pearson Education, 2002.

[Lee99] E. Lee, "Heterogeneous Concurrent Modeling and Design", Superseded by "Overview of the Ptolemy Project," UCB-ERL No. M99/40, July 19, 1999.

[Liu01] Jie Liu, "Responsible Frameworks for Heterogeneous Modeling and Design of Embedded Systems", theses report, 2001.

[Lut00] G. Luttgen, M. Von Der Beek, R. Cleavland, "A Compositional Approach to Statecharts Semantics", 8th ACM SIGSOFT International Symposium on Foundations of Software Engineering, San Diego, California, pp 120–129, November 2000.

[Mad97] "Shore Data Language", Reference Manual, Madison, WI Version 1.1, August 9, 1997.

[Mar91] F. Maraninchi, "The Argos Language: Graphical Representation of Automata and Description of Reactive Systems", in Proceedings of IEEE Workshop on Visual Languages, Kobe, Japan, October 1991.

[Naj99] W. A. Najjar, E. A. Lee, G. R. Gao, "Advances in the Dataflow Computational Model", CAPSL technical Memo, 29, April 1, 1999.

[QoS99] Service en ligne, "Sécurisation des liens d'interconnexion", http://www.arttelecom.fr/dossiers/spectech/4-99_1-0.doc

[Sgr00] M. Sgroi, L. Lavagno, A. Sangiovanni, "Formal Models for Embedded System Design", IEEE design & test computers, April–June 2000.

[Sim98] D. Sima, T. Fountain, P. Kacsuk, "Advanced Computer Architectures a Design Space Approach", Biddles Ltd., Guildfort and King's Lynn edition, 1998.

[SyC02] SystemC Design language, disponible en ligne, http://www.systemc.org, 2002.

[Syn05] Définition: "Modèle Synchrone et modèle asynchrone", service en ligne, http://searchsmb.techtarget.com/sDefinition/0,290660,sid44_gci213080,00.html.

[Tsi03] A. Tsikhanovich, E. M. Aboulhamid, G. Bois, "Object-Oriented Techniques in Hardware Modeling Using SystemC", NEWCAS, Montreal, Canada 2003, June 17–20.

[Vhd93] "IEEE Standard VHDL Language Reference Manual", IEEE, 1993, STD 1076–1993.

Chapter 3

Execution Models

L. Kriaa
Tima Laboratory, Grenoble, France

1. Introduction

A system consists of a set of interconnected subsystems. Each subsystem can have its own characteristics (functionalities and implementations). The computation models as seen in the previous chapter concerning system functionalities, make it possible to define certain features such as parallelism, type of information handled.

However, when dealing with the final execution of the system it is not sufficient to study only these features. In fact, the hardware or software implementation of the subsystems will strongly influence such features. For instance, the parallel execution of a set of operations may be directly implemented into the hardware, as a combinatorial circuit, whereas in software, one has to rely on the scheduler of an operating system to manage the execution of a set of concurrent tasks.

When designing a system, its implementation must be analyzed. The implementation defines the execution model. It may be hardware, software, or functional. These models are described in detail in Section 2. They will be described according to their different levels of abstraction.

The execution of a system is not limited to its behavior, it also includes interconnections between different subsystems. For instance, VHDL interconnections between the various subsystems are defined by sets of signals as well as data propagated through these signals. In C, however, interconnections may be defined as function calls between subsystems. These calls will be interpreted during the compilation and linking phases to implement the global behavior. The study of interconnection execution will be detailed in Section 3. Interconnection implementation is defined at several levels of abstraction. Certain

G. Nicolescu and A.A. Jerraya (Eds.), Global Specification and Validation of Embedded Systems – Integrating Heterogeneous Components, 31–46.

execution languages which are used to implement these interconnections are presented and classified in Section 4.

2. Execution Model of a System

An execution model specifies the way computations occur, which may be of different natures: software, hardware, functional or mixed.

An execution model can be viewed as the interpretation of a computation model. Thus, each computation model can be associated with one or several execution models. A system may be implemented in a homogeneous way (i.e. exclusively in hardware, software, or functional). An execution model is also characterized by an abstraction level that defines the granularity of the communication operations used by the module. It turns out that the unit used to measure granularity may differ according to the domain in which the module is realized. Granularity depends upon the refinement techniques used to move from one abstraction level to another. Hence, for software, computation operations (instructions) and input/output operations are most relevant. Instructions allow for the abstraction of the underlying machine on which software is executed, including input/output operations (interconnection with the external environment).

Software refinement thus consists of converting both computation operations into more basic instructions and input/output operations into corresponding adaptation layers.

For hardware, time representations (continuous, clock cycle, transaction, message, and service) are of the utmost importance since the refining hardware consists of defining a temporal machine to execute functions. For functional modules, the abstraction concerns data representation (abstract data types, bits, etc.) and interaction modes between functions.

2.1 Execution Models of Components

A component execution model represents the realization of the module functionality. This realization can be of three types: software, hardware, or functional.

Based on the concept of communication and computation orthogonality, the execution model of a component is defined as follows:

- A section that implements computation, called basic computational execution. Computational execution requires a suitable engine which interprets a model of computation.

- A section devoted to communication, called the "communication interface", which is responsible for communication between the components (modules) and the rest of the system. This interface is defined by a set of

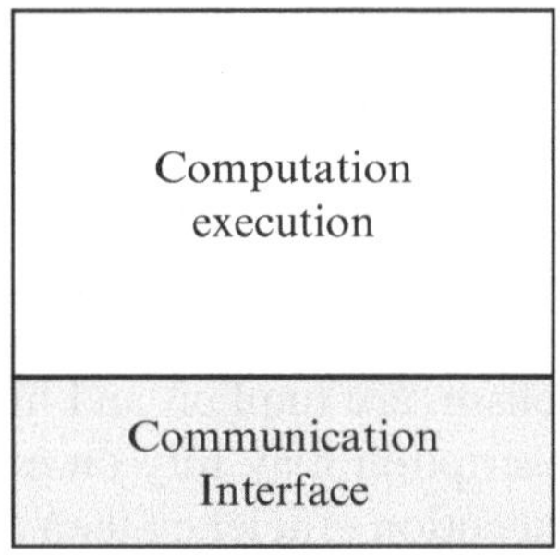

Figure 3.1. Execution model of a component

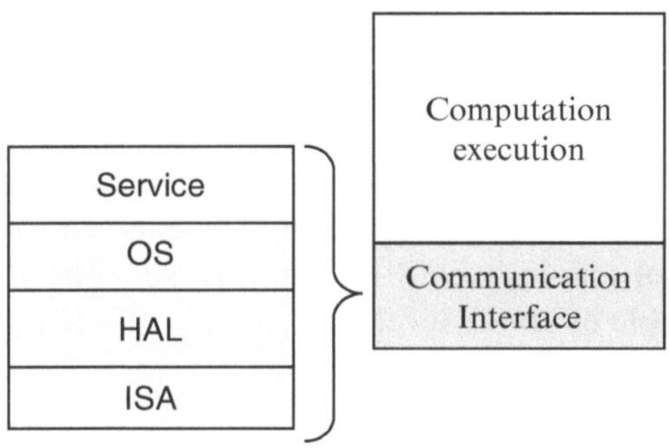

Figure 3.2. Abstraction levels of the communication interface of the software execution model

communication services (provided or required). Different types of communication interfaces may be identified for each execution model.

Figure 3.1 depicts an execution model of a component.

The remainder of the chapter presents the execution models and their respective characteristics.

Software execution model. For software, behavior execution is performed through the interpretation of a set of instructions. The software execution model relies on a hardware platform that interprets a given program (a set of instructions).

The specifications of the hardware platform depend upon the granularity of the interpretation process. Such granularity is related to computational abstraction (data operations). There are four levels of granularity (see Figure 3.2): the service (or application) level, the operating system level (OS), the hardware abstraction layer (HAL), and the instruction-set architecture level (ISA) [Tan99]. Each one of these levels will be addressed individually.

Service level. The service (or application) level is defined as an abstract interpretation of software communication (either between the different applications or between the module and the remainder of the system). At this level, the programmer totally abstracts the underlying machine executing the program as communication may be specified independently from the actual execution machine. Implementation details regarding communication, such as synchronization and parallelism, are implicit and hidden by the API provided at this level. The model description may rely on executable programming or modeling languages. The execution efficiency then depends upon the power of the underlying compiler or interpreter. CORBA [Gei97] and DCOM [Dco05] are examples of typical descriptions at this level.

OS level. At this level, the API is limited to services provided by a given operating system. The use of an operating system makes task synchronization explicit. Communication protocols used between tasks and the external environment are restricted by the operating systems. However, driver implementations (channel topology etc.) remain abstract. The POSIX thread API [But97] is an example of an OS level API.

HAL level. At this level, the communication interface of the software execution model is viewed as a set of communication services which make possible the abstraction of the physical access to hardware. The HAL API is composed of a set of primitives which abstract the basic functions of hardware access such as initialization (boot), context switch or basic input/output operations, on peripherals. The RISC OS HAL API [HAL02] represents an example of this level.

ISA level. The ISA level corresponds to software specifications in assembly language relative to a targeted processor. All communication details are made explicit (register transfers, interruptions, etc.). Indeed, at this level, all hardware details are fixed, and hardware function implementations are thus known.

Hardware execution model. In hardware execution models, it is not necessary to interpret transformation functions to achieve a given specification. Instead, direct mapping of the specification on the hardware configuration is performed.

The execution of the hardware module behavior corresponds to computations realization, which consist of a wired execution. For example, executing an addition (e.g. $a + b$) involving hardware requires an adder circuit.

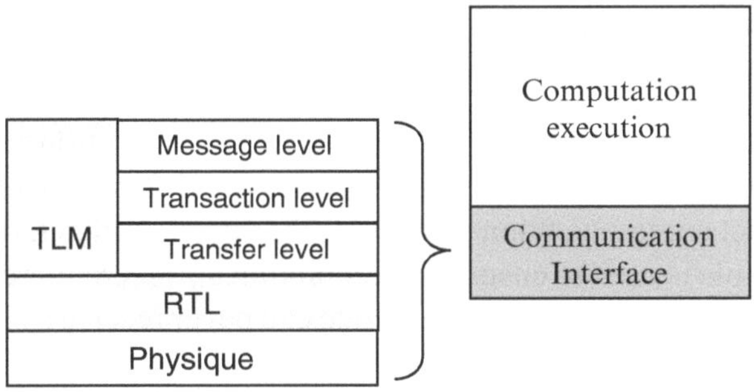

Figure 3.3. Abstraction levels of the communication interface of the hardware execution model

For hardware, the abstraction level concept is generally based on temporal abstractions [Bur03; Con03]. There are three temporal levels (see Figure 3.3): (1) the physical level, (2) the register transfer level (RTL), and (3) the transaction level modeling (TLM).

Time can be modeled at different abstraction levels. Here are more details concerning each of these levels:

The physical level: As seen in electrical level simulations, time values are considered real numbers. Modeling is based on transistors and time is considered a continuous and indivisible unit.

The register transfer level (RTL): The time unit is indicated by an integer (e.g. clock cycle). Nevertheless, communication delays related to logic gates can be associated with this level. The data exchanged are defined at the bit level.

Transaction level: At this level, the model is independent from lower level implementation details such as bus data size or types of used protocols. Some parameters are abstracted, including data types and time of exchanges, by using abstract data types and logical clocks respectively. According to these parameters, a sublevel decomposition may be performed including the transfer, the transaction and the message levels. Time notion thus becomes related to the order in which the transactions occur.

- **Transfer layer:** Time units are still associated with clock cycles. However, certain information, such as memory organization, remains hidden at this level. Transferred data are abstracted through more abstract types of data (e.g. integers).

- **Transaction layer:** A time unit corresponds to a transaction within the selected communication protocol. The abstraction concerns transaction details between the communication bus and memory for instance. Control signals (specific to the protocol selected) are used to express the order in which transactions occur. A single transaction may correspond to several clock cycles. For instance, a read operation for a single word may take either one cycle or four successive cycles.

- **Message layer:** This layer corresponds to the abstraction of the interconnected structure (or architecture) between the different hardware blocks: an abstract communication protocol is then used. At this level, the causality principle is used to ensure that communication remains coherent. This principle can simply be stated as: "one can never receive a message which has not been sent".

Functional execution model. The functional model is independent from the final realization of the component (hardware or software). This model can be compared to a simulated hardware model or to a software model that is natively executed on the host workstation. Within a functional model, component functionalities and their simulation models are most relevant.

The functional execution model corresponds to a virtual execution of the system using executable modeling languages in their simulation environment. The execution of the behavior relies on modeling language features which are relative to language syntax and semantics.

The communication interface is defined as a set of services which allows communication between different modules. These services are generally expressed by the primitives of the description language used.

The functional model deals primarily with data processing itself, rather than the way this processing is eventually realized or implemented. Using this model, only dependencies and relationships between values and functions are expressed. Abstraction levels of the functional model depend on the interaction mode between the different functions. There are four abstraction levels for the functional execution model.

The service level. At this level, the interaction between the functional components and the external environment is performed through a high level API that defines a set of requirements and communication services to be provided. Provided services correspond to API components that are made available to other components. Required services represent calls to functions provided by other components. Dependencies between values and functions are expressed through service-calling mechanisms. These mechanisms are not specifically assigned to a predefined component, but rather, to components able to provide the service. A well-known execution model used at this level is the CORBA model [Gei97].

Message level. At this level, as compared to the service level, the interaction between components is more refined. Moreover, it is defined by a set of API which provides services for sending and receiving messages. Data dependency

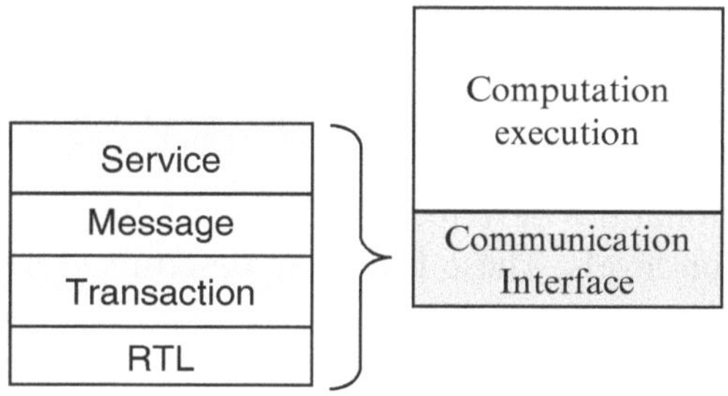

Figure 3.4. Abstraction levels of the communication interface of a functional execution model

is made explicit through service arguments. For instance, using message passing interface (MPI), a component is able to specify the destination of a message. The most commonly used languages which allow for such descriptions are the system description languages (SDL) [Sdl87] and the MPI [Gro96].

Transaction level. At transaction level, an interaction is defined by a particular mode (for instance, sending a message requires two subactions: a request sending and an acknowledgment which notifies the end of the transaction). This model depends on the specific communication protocol used by the components. At this level, the interface is defined as a set of logic ports (see Chapter 2). Data dependency is defined by the selected protocol. CSP [Sch99], behavioral SystemC [SysC02], behavioral VHDL [Vhdl93], Cossap [Cos97] and StateCharts [Har87; Lut00] are examples of languages which allow system descriptions at this level.

RTL (register transfer level). At this level, the ordering of the interactions is performed by a timing unit defined by the underlying language. The interface is defined as a set of physical ports (see Chapter 2). Data dependencies are expressed by sending or receiving a signal transmitted via ports. Manipulated data are considered at the bit level. A number of languages allow component description at this level, e.g. assembly languages, RTL SystemC, RTL, VHDL.

Figure 3.4 illustrates the functional execution model.

3. Execution Model of System Interconnection

This section addresses distributed systems, which are sets of interconnected components. Execution models for interconnections allow for the definition of the way communication is interpreted (executed). This describes how the main functions of a model of computation can be realized (i.e. data transfer, conflict management related to synchronization of modules and quality of service control through scheduling policies).

The interconnection execution model may range from the interpretation of a simple physical wire without underlying behaviors and simple interfaces to more complex models (such as for buses or networks) characterized by complex behaviors and interfaces.

The three functionalities described in Chapter 2, Section 5 can be viewed at different abstraction levels. These levels are determined with respect to well-defined criteria.

For data transfer, the criteria concern the type of media and the nature of the data transferred. For abstraction levels are defined: the service level, the transaction level, the transfer level and the physical level.

There may be different types of synchronization depending on the underlying time model (use of a clock vs. the causality principle).

For control, abstraction levels are closely linked to data transfer levels. In fact, depending on the underlying media and the data used, control may be more or less complex. The abstraction levels of control are the service level, the transaction level, the transfer level, and the physical level.

Since the three functionalities are closely linked, the number of possible abstraction level combinations becomes reduced. In the remainder of this section, the different abstractions levels of each functionality will be considered separately.

3.1 Abstraction Levels of Data Transfer

Data transfer abstraction is based on two criteria: the nature of the transmission media and the type of data manipulated [Nic01]. These are the four abstraction levels:

The service level. At this level, interconnection is abstract (see Chapter 2). The communication media is defined through the interaction mode between components. The interaction mode is carried out via service invocations (function calls). The media is defined as an abstract communication bus such as the object request broker (ORB) bus of CORBA standard [Cor04]. The type of data transferred is not necessarily fixed and corresponds rather to a type of abstract data.

The transaction level. At this level, the interconnection media is represented by a logical network which is an abstraction of a physical interconnection (e.g. wires or physical buses). It may be viewed as a logical association of physical connections. For instance, a FIFO access port is an abstraction of three physical ports (two for control signals and one for data). The interaction mode is defined as a transaction when data transfers are carried out via the media. The types of data transferred are either scalar or complex data structure.

The transfer level. At this level, the media corresponds to a physical bus where some details have been abstracted. For instance, an AMBA bus model described at the transfer level does not model the physical propagation delay of data, but rather a transaction that may correspond to several cycles. This type of data may be bit or bit vector.

The physical level. At the physical level, the media corresponds to a set of wires or physical buses. At this level, all data are bits having a fixed representation. Time can either be discrete or continuous.

3.2 Synchronization Modes

Synchronization at the interconnection level defines how data transfers are carried out. Synchronization is defined according to two time modes: physical or logical.

Synchronization is defined according to:

- A temporal unit defined in the underlying model, that is, for each event observed in this unit (e.g. positive or negative clock edge), data transfers between components are carried out. This is referred to as synchronization on physical time.

- Data sent and received, when a logical clock is used. The logical clock is defined with respect to the "causality principle". The transmission mode is synchronous when the sender and the receiver must synchronize for data transfers (e.g. in socket-based communication [Soc05]). This corresponds to a synchronous logic time mode. In all other cases, the sender and the receiver do not need to be synchronized. An interlocutor can send data without confirmation that the data was received by the other party. This corresponds to an asynchronous transfer mode. FIFO-based communication is an example of this mode.

3.3 Interconnection Control Abstraction Levels

Control is related to behavior (or protocol). This includes scheduling and arbitration functionalities (some resolution functions). Scheduling refers to a set of decisions aimed at establishing an order of application for a set of actions, prior to their execution [Dic05]. Arbitration refers to the process of choosing one of the conflicting demands in order to access a shared resource. Control abstraction levels are established according to the modeling of these functionalities.

The service level. From a user's point of view, interconnection control is transparent at the service level. However, to ensure appropriate behavior, arbitration, and scheduling functionalities become complex. In fact, to enable such transparency for the user, the interconnection model should take into account the information exchanged between the different components, with non-specific data dependencies. For example, a CORBA component may require a service S without knowing the identity of the provider of this service. Therefore, through "ORB" interconnections, the CORBA component requests a service from the remainder of the system components. The ORB is then responsible for dispatching this request to the components concerned and returning the result to the requesting component [Gei97].

The transaction level. At the transaction level, interconnection control is defined through the scheduling of different data transactions (sending or receiving) between components. At this level, data dependency between components is set by the implemented protocol by the media (e.g. a blocking FIFO). In this case, scheduling is responsible for ensuring transaction routing as defined by the underlying protocol. For instance, for the blocking FIFO, read access is granted to all components connected to it as long as it is not empty.

The transfer level. Interconnection control, at the transfer level, corresponds to data transfer scheduling according to the operating mode of the interconnected bus e.g through bus controllers (like AMBA bus controller [Ahb99]).

The physical level. At the physical level, interconnection control is ensured by resolution functions of physical signals. At this level, all interconnections and signal routings are defined by a physical topology.

Summary. Table 3.1 shows interconnection features throughout the different abstraction levels.

The study of various modeling and specification languages allowed us to establish a classification according to the interconnection modeling capabilities of these languages, at the different abstraction levels previously discussed.

Table 3.1. Basic interconnection features at different abstraction levels

Abstraction Level	Communication Media	Behavior	Data Type
Service Level	Abstract media (e.g. ORB)	Function calls	# uninterpreted
Transaction Level	Logical channels (e.g. FIFO)	Abstract protocols	Scalar/date structure
Transfer Level	Explicit channels (e.g. Bus)	Fixed protocol	Bits or bit vectors
Physical Level	Physical wires	Signal transfer	Signals

Generally, modeling languages define interconnections as components of their execution semantics. Some languages include a unique definition of interconnection, such as the CORBA remote function call.

However, other languages can be used when defining several abstraction levels for interconnection. These include the major part of hardware description languages such as SystemC or VHDL. For instance, SystemC can be used to model interconnections at the physical level which correspond to a physical wire or signal (the role of the *sc_signal* primitive). It can also be used to describe interconnections at the transaction level using a FIFO channel for instance (*sc_fifo* primitive). These features can be defined with respect to the interconnection execution models on which these languages are based. These models will be covered in Section 4. Hardware description languages (HDL) are based on a representational model that is able to cover a large realization spectrum throughout the different abstraction levels.

Interconnection realization is viewed according to three axes, which correspond to three functionalities: synchronization, interconnection control, and data transfer.

However, for each language, an execution model of the interconnection is specified. In the following section, this aspect will be considered and corresponding models will be associated with each language.

4. Classification of Current Interconnection Execution Models

Interconnection descriptions are achieved through the so-called execution models. These models can be represented at different abstraction levels according to the more or less detailed descriptions of some properties related to communication (channel, protocol, etc). Furthermore, the same models can also be classified according to the way synchronization is achieved in order to carry out communication. In the following section, taxonomies for the most commonly used interconnection execution models are presented.

4.1 Existing Interconnection Execution Models

Several execution models exist to perform interconnections. This paragraph presents some of the most popular:

- Remote invocation models (RI)
- Synchronous message passing models (SMP)
- Asynchronous message passing models (AMP)
- Synchronous models
- Discrete event models (DE)

Remote invocation models (RI). The remote invocation communication model is based on calls to remote services. There are several remote invocation models. Some examples include the client/server model, the remote procedure call (RPC) model, etc.

The client/server model describes the relationships between two programs which may be written in different languages. One program, called the client, requires services from another program; the other program, called the server, is in charge of providing services to the client. This model is particularly well-suited for connecting distributed programs.

The RPC model has been defined by Birell and Nelson [Rpc05]. It represents a protocol that requires the execution of distant procedures. For example, Open Network Computing Remote Procedure (ONC RPC) and CORBA are RPC-based protocols.

RPC models can be classified as synchronous or asynchronous.

- **Synchronous RPC models:** The client is blocked while waiting for the server's response (as in regular local procedure calls). This model is intuitive and allows for the swift detection of errors. Furthermore, it does not require data buffering.

- **Asynchronous RPC models:** The client is not blocked while waiting for a response from the servers. Instead, an acknowledgment process allows the notification of the server response. This model requires data buffering to save responses coming from the server. Moreover, communication must be reliable; this implies the use of lossless channels. The X11 graphic server and file printing are two examples of systems relying on such a model.

Synchronous message passing (SMP). The SMP model relies on the passing of messages to ensure component synchronization. In this model, synchronization, data conversion, and control management are explicit. The communication is then carried out in an atomic way. Components, which are processes, communicate using a "rendez-vous" protocol which makes the identification of a global system state possible. The SMP model is mainly used in systems which make an extensive use of shared resources [Kai02]. However, since this model exhibits strong coupling between components, maintaining the entire system, and enforcing determinism become very tedious tasks. CSP [Sch99] and Occam [Lee99] are two languages that support SMP-like modeling.

Asynchronous message passing (AMP). The AMP model relies on the concept of process network (PN) allowing the representation of data flow based behaviors [Lee01; Lee99].

In this model, communication is performed via messages sent through communication channels. There is no temporal dependency between components, which prevent implicit acknowledgment between components-related processes [Kai02].

With this model, hardware and software implementations are relatively simple. It refers to the flexibility with which different commands can be transferred. Furthermore, the model is particularly efficient with signal processing system cases. SDL [Sdl87], Promela [Hol90], and Estelle [Est05] are examples of languages which support this model.

Synchronous models. Synchronous systems are based on concepts of global clock, immediate information broadcast, deterministic parallelism and preemption to provide a coherent and suited programming model.

This model supposes that all modules are cadenced by the same clock (physical and/or logical). Each signal can be associated with a clock (either explicit or virtual). This clock has a relative meaning compared to other signals associated with clocks which allow establishing a total time order.

Every instant, a synchronous system executes a transition using memorized signals. To determine the state of any given system, two phases are carried out: the first consists of memorizing the signal and the second the execution of a transition of all processes with memorized signals. The interaction between components, which can be expressed as relations between input and output signals, is then performed through clock-aligned data.

This model is particularly well adapted for application with a high concurrency level and a complex logical control [Lee99]. Communication between the different modules is performed using data values which are aligned on clock edges. For example, Esterel and SyncCharts [And98] are two languages which allow descriptions based on such a model.

Discrete event (DE) models. In the discrete event model, time and state vector elements are discrete variables. This model corresponds to a totally ordered model and makes use of an explicit time representation, making it suitable for both simulation and implementation [Lee99].

Communication is carried out through events belonging to the same time ax (in a discrete way) thanks to read/write primitives on shared signals. When executed, components consume input events and produce output events. This generation process respects the causality principle which states that output events cannot be generated prior to input events.

The discrete event model is particularly well-adapted for hardware descriptions as it expresses hardware concurrency in a natural manner. Modeling languages supporting such a paradigm include VHDL and SystemC.

Table 3.2. Classification of interconnection execution models through levels of abstraction

Abstraction Level \ Synchronization	Synchronous	Asynchronous
Service Level	Synchronous remote invocation	Asynchronous remote invocation
Transaction Level	Synchronous message passing	Asynchronous message passing
Transfer/Physical Levels	Synchronous	Discrete Event

Classification of current interconnection execution models. In this section, the execution models previously described are classified according to the abstraction levels described in Section 3. Table 3.2 illustrates this classification.

However, such a classification remains limited. Indeed, given the large number of models commonly used for communication, establishing an exhaustive represents a major challenge. We have thus limited our study to a subset of these models.

5. Conclusion

This chapter presented different execution models of a system. These models may be designed for software, hardware or functional purposes. All of these models can be described at several levels of abstraction.

Execution models of system interconnections were also presented. These may be seen as realizations of associations between different communication interfaces of different components. These models are defined at different levels of abstraction.

The classification of current interconnection execution models, according to the synchronization mode used for data transfer between components, was introduced.

However, defining a global execution model for heterogeneous systems remains challenging. In fact, having a global model that encompasses all specific subsystem models along with the interconnection model requires the principle of heterogeneous interconnection of heterogeneous systems due to the manipulation of different concepts and properties.

References

[Ahb99] *AMBA Specification*, ARM Limited Edition, 1999, available online http://www.gaisler.com/doc/amba.pdf.

[And98] C. André, H. Boufaied, S. Dissoubray, "SyncCharts/Esterel: un modèle synchrone pour systèmes réactives complexes", in *Real-Time and Embedded Systems*, Paris, 14–16 January 1998, pp. 175–194, Teknea.

[Bur03] M. Burton, F. Ghenessia, Stuart Swan, "Transaction Level Modeling: Overview and Requirements for SystemC Methodology", White paper, May 13, 2003.

[But97] D. R. Butenhof, *Programming with Posix Thread*, Addison-Wesley Professional, May 16, 1997.

[Con03] J. Connell, "ARM System-Level Modeling", White paper, June 25, 2003, www.arm.com.

[Cor04] Available online, "Common Object Request Broker Architecture: Core specification", v3.03.3, March 2004, OMG, Inc. http://www.omg.org/technology/documents/corba_spec_catalog.htm.

[Cos97] Synopsys, Inc., 700 E. Middlefield Rd., Mountain View, CA 94043, USA, COSSAP User's Manual, January 1997.

[Dco05] DCOM, Definition online, http://www.webopedia.com/TERM/D/DCOM.html.

[Dic05] Scheduling, Definition on line, http://w3.granddictionnaire.com/btml/fra/r_motclef/index1024_1.asp

[Est05] Estelle, service online, http://www.infres.enst.fr/~cottin/protocol/estelle.htm.

[Gei97] J. M. Geib, C. Gransart, P. Merle, *Corba, des concepts à la pratique*, Masson Editeur, Paris, 1997.

[Gro96] W. Gropp, E. Lusk, A. Skjellum, *Using MPI: Portable Parallel Programming with the Message-Passing Interface*, MIT Press, 1996.

[HAL02] On line service, "Hardware Abstraction Layer in RISC OS 5", December 2002, http://www.iyonix.com/32bit/HAL.shtml.

[Har87] D. Harel, "A Visual Formalism for Complex Systems", Science of Computer Programming, 1987, 8, pp 231–274.

[Hol90] G. J. Holzmann, *Basic Spin Manual*, 10th Edition, Vol. 2, Saunders College, 1990, pp. 429–450.

[Kai02] C. Kaisar, "Synchronization Par Message", course support, May 2002.

[Lee01] E. Lee, "Computing for Embedded Systems", IEEE Instrumentation and Measurement Technology Conference Budapest, Hungary, May 21–23, 2001.

[Lee99] E. Lee, "Heterogeneous Concurrent Modeling and Design", Superseded by "Overview of the Ptolemy Project," UCB-ERL No. M99/40, July 19, 1999.

[Lut00] G. Luttgen, M. Von Der Beek, R. Cleavland, "A Compositional Approach to Statecharts Semantics", ICASE Report No. 2000-12, March 2000.

[Nic01] G. Nicolescu, K. Svarstad, W. Césario, L. Gauthier, D. Lyonnard, S. Yoo, P. Coste, A. A. Jerraya, "Desiderata pour la spécification et la conception des systèmes électroniques", Technique et science informatiques, 8 avril 2001.

[Rpc05] Available online, http://www.faqs.org/rfcs/rfc1831.html.

[Sch99] S. Schneider, *Concurrent and Real-Time Systems: The CSP Approach*, Wiley Professional Computing Edition, September 1999.

[Sdl87] Computer networks and ISDN Systems. CCITT SDL, 1987.

[Soc05] Sockets, service on line, http://www.commentcamarche.net/sockets/sockintro.php3

[SyC02] "SystemC Design Language", available online, http://www.systemc.org, 2002.

[Tan99] Andrew S. Tanenbaum, *Structured Computer Organization*, Prentice-Hall, 4th Edition, 1999.

[Vhd93] *IEEE Standard VHDL Language Reference Manual*, IEEE, Institute of Electrical and Electronically Engineers, 1993, STD 1076-1993.

Chapter 4

Heterogeneous Systems Validation Based on Execution Models

G. Nicolescu[1] and A.A. Jerraya[2]
[1]*Ecole Polytechnique de Montreal, QC, Canada,* [2]*CEA-LETI, Grenoble, France*

1. Introduction

Currently, the validation stage represents a bottleneck for the design flow of heterogeneous systems. This stage requires most of the design time (70–80%) [Kea99] and this is mainly because the different components of an embedded system are heterogeneous. This characteristic of current embedded systems makes very difficult the definition and implementation of global execution models enabling the global validation. These models are very complex: they guarantee the execution of different components, the interpretation of interconnections between different components and they are in charge of adaptations between different execution models and/or communication protocols. Consequently, the development of these models is a time consuming task; moreover, often it represents a source of errors during the design flow. In this context, the automatic generation of global execution models for global validation purposes becomes mandatory.

This chapter focuses on the definition of the global execution models of heterogeneous systems. It introduces the basics on global execution models and presents the main characteristics required for these models in order to facilitate their automatic generation and the efficient global validation. The existing works proposing validation for heterogeneous systems based on global execution models are also presented.

2. Global Execution of Heterogeneous Systems – Basics

The global execution of heterogeneous systems may be achieved using two main approaches:

- Compositional approach, where graphical or multilanguage specification of a heterogeneous system are translated into a specification using

G. Nicolescu and A.A. Jerraya, Global Specification and Validation of Embedded Systems – Integrating Heterogeneous Components, 47–57.

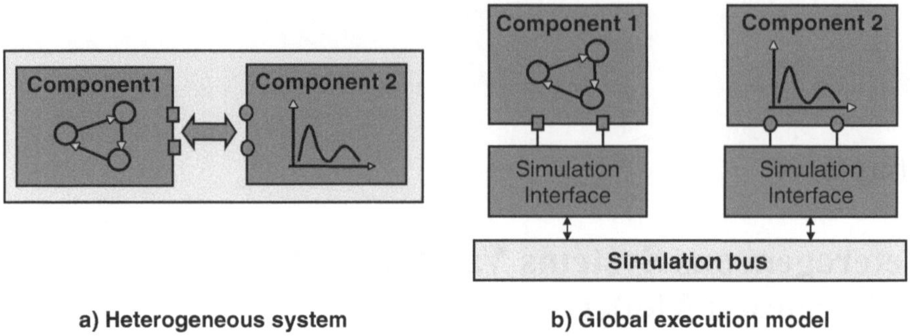

Figure 4.1. Heterogeneous system and its corresponding global execution model

a unique language. The resulting specification is then executed for the global validation [Cal96; Pas99].

- Distributed approach, where the execution of a heterogeneous system is realized by the joint execution of different simulators. In this case, the initial specification for each component of the system remains unchanged.

In this chapter we will focus mainly on the distributed approach.

A heterogeneous systems and its corresponding global execution model are illustrated in Figure 4.1. As shown in the figure, three types of basic elements compose the global execution model: the execution models of the different components constituting the heterogeneous system, the simulation bus and the simulation interfaces. The remaining of this paragraph will define the simulation bus and the simulation interfaces.

2.1 Simulation Bus

The simulation bus is in charge of interpreting the interconnections between the different components of the system. The simulation bus must be able to interpret the different types of interconnection:

- Service level interconnections are represented by a combination of requests and services. The different components communicate by services requests, via abstract network guaranteeing the routing and the synchronization of communication that is established dynamically. The typical communication primitive is a service request such as "print(file)". Common object request broker architecture (CORBA) [Omg97] is a typical example of this abstraction level.

- Transaction level interconnections enabling the different components of the system to communicate through an explicit network composed of communication channels, named active channels. In addition to the synchronization, these channels may contain a complex behavior such as the conversion of the protocols specific to the different communicating modules. The communication details are encapsulated by high level communication primitives (e.g. send/receive) and no hypothesis on the realization of the communication is made. System description language (SDL) [Sdl87], unified modeling language (UML) [Uml05], and ObjecTime [Obj05] are examples of languages modeling concepts specific to this abstraction level. Certain implementations of StateCharts [Har87] may be placed at this abstraction level.

- Transfer level interconnections are abstract and they encapsulate fixed protocols. Consequently, at this level, a model implies the choice of the communication protocol. Typical examples of communication primitives are writing a value or waiting for an event. Languages modeling systems at this abstraction levels are CSP [Hoa85], SystemC [Sys00], and StateCharts [Har87].
- When the interconnections are at the physical level, the communication is realized by physical wires and buses. The granularity of the unit time is the clock cycle and the communication primitives are set/reset on ports and waiting a new clock cycle. SystemC [Sys00], Verilog [Moo98], and VHDL [Iee93] are the most used languages to describe systems at this abstraction level.

2.2 Simulation Interfaces

The simulation interfaces enable the communication of the different components of the system through the simulation bus. Their role is to adapt each component to the simulation bus. Consequently, they are in charge of:

- Adaptation of different execution models – in order to enable the global execution of systems including components presenting different execution models (e.g. components at different abstraction levels or components designed using different technologies like hardware/software or continuous/discrete). In this case, the simulation interface has to provide refinements for communication primitives and/or different synchronization mechanisms
- Adaptation of different simulators to the simulation bus – in order to guarantee the transmission of information between simulators executing the different components of the heterogeneous systems
- Adaptation of different communication protocols – in order to enable the global execution of systems including components presenting different communication protocols.

2.3 Simulation Backplane

The simulation backplane is the element of the global execution model that guarantees the synchronization and the communication between the different components of the system. It is composed of the simulation interfaces and the simulation bus.

2.4 Simulation Instance

The implementation and the simulation of an execution model in a given context is called simulation instance. Several instances may correspond to the same execution model and these instances may use different simulators and may present different characteristics (e.g. accuracy, performances).

2.5 Simulation Environment

The simulation environment is an instrument for the creation of the simulation instances. In case of the global validation of heterogeneous systems, the simulation environment must be able to receive as input a heterogeneous specification and to generate automatically the simulation bus and the simulation interfaces required for the global execution of the system.

This type of environment requires a conceptual foundation defining the different basic elements and the algorithms that are used. Two important aspects of a simulation environment conceptual basis are:

- The definition of the execution model – the definition of the simulation bus and simulation interfaces

- The definition of the methodology for the automatic generation of simulation instances

These two aspects are related: the definition of the simulation bus and simulation interfaces may facilitate the automatic generation of simulation instances. The latter may influence the quality of the simulation interface and simulation bus. These two definitions have direct consequences on the efficiency of the simulation environment. In the next paragraph, the main characteristics required for an execution model targeted by a simulation environment are presented.

3. Required Characteristics for Global Execution Models of Heterogeneous Systems

The main characteristics required for the execution models in order to facilitate their automatic generation and to provide the efficient global validation are: (1) flexibility, (2) scalability and modularity, and (3) accuracy.

3.1 Flexibility

A flexible execution model is able to be adapted to the modifications occurring during the design flow:

- The modification of the external environment (e.g. the modification of the functional constraints)
- The modification of the utilization mode (e.g. the modification of the communication protocols)
- The technology modification (e.g. the replacement of a given simulator by an other)

3.2 Modularity and Scalability

An execution model must be also modular and scalable. Thus, it must enable the independent handling of the different components of the systems. It must enable also the evolution of the system complexity.

The main advantages of using such an execution model are:

- The possibility to validate the different solutions for the realization of a given system
- The possibility to validate a system when new components or new functionalities have to be added
- The validation of systems and their environment

3.3 Accuracy

Depending on the accuracy level, may two types of validation may be defined:

- The timed validation taking into account the time notion and enabling to evaluate the performances and the time constraints of a system.
- The functional validation enabling the validation of the functionality of a system. In this case, only the data transfers are considered without taking into account the time required for their transfer.

An efficient execution model must allow designers the choice of the accuracy level.

4. State of the Art on Heterogeneous Systems Validation

Several environments and methodologies for global heterogeneous systems were already proposed. Most of these works propose the validation at register transfer level (RTL), presented in Chapter 3. Some works present methodologies for integrating several specification languages and simulators for global validation of systems composed of subsystems from different application domains. Recently, new techniques are focused on mixed-level simulation.

The main works for each of these directions will be presented in the following.

4.1 Tools for Hardware/Software Simulation

Currently, the simulation is very popular in industry for the global validation of hardware/software systems at RT Level. The main objective is to validate the hardware and software integration into a virtual framework, before the design of the first prototype.

The RTL hardware/software simulation enables the joint simulation for hardware and software components. The software part of the system is simulated using one or several instruction set simulators (ISSs) and the hardware part of the system is simulated on a hardware simulator (e.g. VHDL, Verilog and more recently SystemC).

Mentor Graphics Seamless CVE [Men02], Synopsys Eaglei [Syn02], and CoWareN2C [Cow02] are commercial tools enabling joint simulation of several ISSs and a hardware simulator.

Albreght et al. [Alb93] presents a methodology for the performance estimation of hardware/software simulation using ISSs and hardware simulators. Synopsys Eaglei [Syn02] is used for their experiments.

In the same context of hardware/software RTL simulation, Ghosh et al. [Gho95] propose an environment in which several simulator may be executed concurrently. Their work is focused on the synchronization between several simulators.

This type of global validation is very interesting for the accuracy that may be provided. Unfortunately, since the hardware/software systems are more and more complex, the performances of these techniques become problematical for the design space exploration.

4.2 Tools for Computing Models Composition

Several existing works on global validation are motivated by the diversity of computing models for the different subsystems composing a heterogeneous system.

A representative contribution in this direction is Ptolemy [Pto02]. This project focuses on the modeling and validation of systems including heterogeneous computation models. The most important contribution of this project is the definition of a conceptual framework enabling the comparison of different computation models [Lee97]. Starting from this framework, a Java-based simulation tool was defined [Eke01]. This tool allows the global simulation for systems integrating several computation models (e.g. the discrete event (DE) model, the finite state model (FSM), the model based on communicating sequential processes (CSP)). The implementation of a computation model is called domain.

Several simulation tools are based on Ptolemy principle. Thus, the hardware/software design tool Polis [Edw97] uses the DE domain of Ptolemy for the RTL simulation. SPW [Spw05] uses also Ptolemy concepts for the global simulation of digital signal processing (DSP) systems.

4.3 Tools for Execution Models Composition – Multilanguage Validation

Several simulation-based validation methodologies are focused on the automatic generation of execution models integrating several specification languages. This works exploit the libraries provided by the simulators in order to communicate with the external environment.

A tool for the automatic generation of execution models integrating C, VHDL, Simulink [Mat00], and SDL/ObjectGeode [Sdl87] is presented in [Lem00] and [Cos99]. The communication and the synchronization between different simulators involved in a simulation instance are implemented using the inter-process communications (IPCs) or sockets [And91].

An approach for fast validation of hardware systems composed of subsystems described using different languages is proposed in [Hen01]. The specification languages integrated in this tool are Verilog, VHDL, and C-based hardware languages. This work exploits the interface foreign language interface (FLI) of Modelsim [Mod05]. All subsystems composing the system to validate are executed on a single environment (Modelsim). This solves the main problem of the classical multilanguage approaches: the overhead of the communication through IPCs or sockets.

The presented works are very efficient for the integration of several simulators into a simulation instance. However, they are not allowing the integration of several abstraction levels or the discrete/continuous simulation. Thus, in the case where this type of adaptations is required, they have to be done manually by the designer.

4.4 Tools for Execution Models Composition – Multilevel Validation

Recently, several activities on global validation of heterogeneous systems are concentrated on the multilevel aspect. These approaches enable the joint simulation of different components at different abstraction levels. Most of them propose the global simulation of heterogeneous systems integrating both components presenting functional execution models and components presenting cycle-accurate execution models. Among these solutions, the bus functional model (BFM) [Sem00] is widely used. It performs transformation between functional memory access and cycle-accurate memory access.

SystemC provides a concept of interface for mixed abstraction level simulation [Sys00]. Mixed level interfaces have to be designed manually, which can be a source of errors and time consuming. Moreover, designers need also to design wrappers for multilanguage simulation. SystemCSV extends SystemC and makes the simulation of all of its abstraction levels of interfaces [Sie01]. Thus, the interconnection of interfaces with different abstraction levels is possible. However, automatic generation of mixed-level simulation interface is not supported.

VSIA [Lem 00] proposes virtual component interface (VCI) and system level interface (SLI) for heterogeneous components interconnection, where several abstraction levels are taken into account. VCI enables only point-to-point and unidirectional connections. VSIA standard presents a design guide (not an automatic design method) to enable multilevel simulation consisting of VCI and SLI by using the wrapper concept.

CoWareN2C (Synopsys) [Cow02] is an environment providing multilevel simulation. The abstractions proposed are: the bus cycle accurate (BCA) level corresponding to the classical RTL and the UnTimed (UT) level where only the functionality of a system is taken into account and the time is abstracted. For the joint simulation of subsystems described at BCA level and subsystems described at UT level, CoWareN2C presents the bus-cycle-accurate-shell (BCASH) concept. It is a wrapper for UT level subsystems and it enables estimation of the time required for the execution of a given functionality.

Concerning the continuous/discrete simulation, in [Mar02] the authors propose a cosimulation environment based on Xyce (a SPICE parallel simulator) and SAVANT (a parallel VHDL simulator). The interfacing is enabled by C++ classes containing methods for signal conversion and data exchange between simulators.

Other approaches propose the utilization of a unique language for the specification of the overall system. Some of these languages may be obtained by extension of well-established languages. Illustrative examples are VHDL-AMS [Iee99], Verilog-AMS [Fre00], and recently SystemC-AMS [Vac03] extending VHDL, Verilog and respectively SystemC for mixed-signal systems design. Modelica [Mod97] is a language for the design of heterogeneus systems. Several commercial simulation tools are based on it (e.g. Dymola, MathModelica). It provides a set of libraries for several application domains (electrical, thermal, etc). However, the concepts of time and discrete events are difficult to manipulate in this language.

5. Conclusion

This chapter defined the global execution models of heterogeneous systems. It introduced the basics on global execution models and their main components and the main characteristics required for these models in order to facilitate their automatic generation and the efficient global validation. The existing works proposing validation for heterogeneous systems based on global execution models were also presented. The analysis of these work showed that more efforts are necessary to define a validation tool being able to generate automatically global execution models mastering all dimensions of the current embedded systems heterogeneity.

References

[Alb93] T. W. Albercht, J. Notbauer, S. Rohringer, “Hw/Sw CoVerification Performance Estimation & Benchmark for a 224 Embedded RISC Core Design”, Proc. Design Automation Conf., pp. 801–811, June 1998.

[And91] G. R. Andrewus, “Concurrent Programming, Principles and Practice”, Benjamin and Cummings (eds), Redwood City, CA, pp. 484–494, 1991.

[Cal96] J. P. Calvez, D. Heller, O. Pasquier, “Uninterpreted Co-Simulation for Performance Evaluation of Hw/Sw Systems”, Proc. CODES/CASHE’96, Pittsburgh, Pennsylvania, USA, 18-20 mars, 1996 pp. 132–139.

[Cho95] P. H. Chou, R. B. Ortega, G. Boriello, “The Chinook Hardware/Software Co-Synthesis System”, Proc. International Symposium on System Synthesis, 1995

[Cos05] Cosimate overview, available at http://www.tni-world.com/ cosimate.asp

[Cos99] P. Coste, F. Hessel, Ph. LeMarrec, et al. “Multilanguage Design of Heterogeneous Systems”, Proc. International Workshop on Hardware-Software Codesign, May 1999.

[Cow02] Coware. Inc., “N2C” available at http://www.coware.com/cowareN2C.html/.

[Edw97] S. Edwards, L. Lavagno, E. A. Lee, A. Sangiovanni-Vincentelli, “Design of Embedded Systems: Formal Models, Validation, and Synthesis”, Proc. IEEE, Vol. 85, No. 3, March 1997.

[Eke01] J. Eker, C. Fong, Jorn W. Janneck, J. Liu, “Design and Simulation of Heterogeneous Control Systems using Ptolemy II,” Proc.

IFAC Conference on New Technologies for Computer Control (NTCC'01), Hong Kong, China, November 2001.

[Fre00] Frey, P. et al., "Verilog-AMS: Mixed-signal simulation and cross domain connect modules", Proc. Behavioral Modeling and Simulation International Workshop, 2000.

[Gho95] A. Ghosh, M. Bershteyn, et al. "A Hardware-Software Co-Simulator for Embedded Systems Design and Debugging", Proc. Asia South Pacific Design Automation Conference, 1995.

[Har87] D. Harel, "Statecharts : A Visual Formalism for Complex Systems", Science of Computer Programming, 1987, 8, pp. 231–274.

[Hen01] Y. Héneault, G. Bois, E. M. Aboulhamid, "A Fast Hardware Co-Specification and Co-Simulation Methodology Integrated in a H/S Co-Design Platform", Proc. 13th International Conference on Microelectronics, Rabat, Morocco, Oct. 29-31, 2001, pp. 249–252.

[Hoa85] C. A. R. Hoare, *Communicating Sequential Processes*, Prentice-Hall, 1985.

[Iee93] Institute of Electrical and Electronically Engineers, *IEEE Standard VHDL Language Reference Manual*, 1993, STD 1076-1993. IEEE.

[Iee99] *IEEE Standard VHDL Analog and Mixed-Signal Extensions*, IEEE Std 1076.1-1999, 23 December 1999.

[Kea99] M. Keating, P. Bricaud, *Reuse Methodology Manual*, Kluwer Academic, 1999.

[Pas99] O. Pasquier, J. P. Calvez, "An Object-Based Executable Model for Simulation of Real-Time Hw/Sw Systems", in Proc. Design Automation and Test in Europe (DATE), 1999.

[Lee97] A. Lee, A. Sangiovanni-Vincentelli, "A Denotational Framework for Comparing Models of Computation", ERL Memorandum UCB/ERL-M97/11, University of California, Berkley, CA 94720, January 1997.

[Lem00] Ph. Lemarrec, "Cosimulation multi-niveux dans un flot de conception multi-language", Thèse de doctorat, INPG, Tima Labratory, June 2000.

[Mar02] Martin D.E et al., "Integrating multiple parallel simulation engines for mixed-technology parallel simulation", Simulation Symposium, 2002.

[Mat00] Mathworks. 2000, Matlab, http://www.mathworks.com.

[Men02] Mentor Graphics, Inc, "Seamless CVE", available at http://www.metorg.com/semless.

[Mod97] Modelica – A unified object-oriented language for physical systems modeling, specifications report, September 1997, version 1.0, www.modelica.org.

[Mod05] Modelsim, available on line to http://www.model.com.

[Moo98] Philip R. Moorby, Donald E. Thomas, *The Verilog Hardware Description Language*, May 1998, Hardcover.

[Obj05] Objectime, available on-line at http://www.objectime.on.ca/, 2005.

[Omg97] Object Management Group, CORBA Services; Common Object Services Specification, Technical Report, OMG, July 1997.

[Pto02] Ptolemy project, available on-line at http://ptolemy.eecs.berkeley.edu, 2002.

[Sdl87] Computer Networks and ISDN Systems. CCITT SDL, 1987.

[Sem00] L. Semeria, A. Ghosh, "Methodology for Hardware/Software Co-verification in C/C++", Proc. Asia South Pacific Design Automation Conference, Jan. 2000.

[Sie01] R. Siegmund, D. Muller, "SystemCSV – An Extension of SystemC for Mixed Multi-Level Communication Modeling and Interface-Based System Design", Proc. Design Automation and Test in Europe, March 2001.

[Syn02] Synopsys, Inc., "Eaglei", available at http://www.synopsys.com/products/hwsw/eagle_ds.html.

[Sys00] Synopsys, Inc., "SystemC, Version 2.0", available at http://www.systemc.org/.

[Spw05] SPW, available on line to www.cadence.com/products/spw.html, 2005.

[Uml05] UML, available on-line at http://www.rational.com/uml/, 2005.

[Vac03] Vachoux A. et al., "Analog and mixed signal modelling with SystemC-AMS", Circuits and Systems, Proc. ISCAS'03.

Chapter 5

Anatomy of a Hardware/Software Execution Model in Heterogeneous Systems

G. Nicolescu[1] and A.A. Jerraya[2]

[1]*Ecole Polytechnique de Montreal, QC, Canada,* [2]*CEA-LETI, Grenoble, France*

1. Introduction

In order to provide embedded systems with the required efficiency and flexibility, software is increasingly being used by designers to implement functionalities. Consequently, embedded systems software components are becoming more and more complex and the use of operating systems (OS) is mandatory to deal with concurrent and inter-task communication. This evolution makes the validation of hardware/software difficult and the use of classical techniques even more problematic.

Nowadays, the most popular software (composed of an application and an OS) validation technique relies upon processor simulators. As explained in the following section, the complexity of the modern embedded software prevents these simulators from delivering an adequate performance. Therefore, the use of native simulations to validate embedded software represents an interesting alternative. Adopting this alternative is challenging as parts of the OS code depend on the target processor that is used in the embedded system and the native simulation requires the implementation of simulation models for the embedded OS. The main requirements for these models are:

- *Coherence*: The OS simulation model must respect the final implementation of the OS.
- *Accuracy*: The OS simulation model must enable an estimation of the final architectural performance of the embedded system.
- *Interoperability*: The OS simulation model must enable hardware/software validations. Consequently, it must enable communication between software and hardware components.

G. Nicolescu and A.A. Jerraya, Global Specification and Validation of Embedded Systems – Integrating Heterogeneous Components, 59–74.

An OS simulation model will be presented in this chapter. To ensure maximal coherence, the main goal is to enable the native simulation of embedded software by reusing processor-independent codes and providing simulation models only for the processor-dependent OS services. To ensure accuracy, the notion of time is taken into account by adding timing annotations to the OS model; to provide hardware/software validation, the OS model is integrated into a global execution model.

The chapter is structured into six different sections. The following section introduces the main techniques used to validate embedded software and presents state-of-the-art OS validation information. The OS simulation model is presented in the third section, and the fourth section addresses the automatic generation of the simulation model. The fifth section illustrates the use of the simulation model for the overall validation of a very high-speed digital subscriber line (VDSL) modem. The final section presents the conclusion.

2. Validation of Software Components

This section introduces different techniques and OS simulation models currently being used to validate embedded software. Current approaches which exploit these models and techniques are also presented.

2.1 Validation Techniques for Embedded Software

Two main techniques are used to simulate software: the instruction set simulator (ISS)-based simulation and the native simulation.

ISS-based simulation. An ISS is a tool that runs on a host machine and mimics the behavior of a program running on a targeted machine. It enables designers to validate their applications even for cases where target machines do not yet exist or are not yet available. Typically, an ISS allows users to examine the internal state of the target machine (e.g. the processor register values) during the execution of each instruction. ISS are indispensable tools to develop processors and software design processes.

The most important quality metrics of an ISS are [Zhu99]:

- *Simulation speed*: especially relevant to develop high performance systems, even more so when real-time simulations are required
- *Compilation speed*: the capacity to bring an application into a simulation state rapidly
- *Traceability*: the ability to collect relevant statistics, such as instruction profiling
- *Scalability*: the ease with which the tool can be extended to handle new target machines and host platforms

- *Interoperability*: the capability of integrating other tools, such as debuggers, hardware simulators, etc.

Most commercial ISSs are interpretation-based ISSs. Interpretation-based simulation builds a data structure in the memory representing the state of the target processor. It then enters a loop in order to execute four main functions:

- Fetch: reads an instruction word from memory.
- Decode: analyzes the fetched instruction and extracts the opcode field for the instruction.
- Dispatch: uses a switch statement to jump to the appropriate code and handle a particular instruction.
- Execute: updates the processor's state according to the semantics of the instruction.

Although their implementation is simple and flexible, interpretive simulators are hindered by performance problems mainly due to the overhead spent in order to fetch, decode, and dispatch instructions, which is unproductive from a simulation perspective [Zhu99]. Most interpretative ISSs execute from 10 K to 100 K of instructions per second [Row94], which is not practical when simulating software composed of thousands of lines of code.

Due to productivity problems for interpretation-based ISSs, several studies focus on defining new techniques for fast hardware/software simulations [Ziv96] and propose compilation-based ISSs.

Compilation-based approaches reduce runtime overhead by translating each target machine's instruction directly into a series of host machine instructions which manipulate the simulated machine's state. For example, during a simulation, an MIPS code which is supposed to run on an MIPS processor in the final implementation of an embedded system is translated into SPARC code (in the cases where the host machine executing the simulation contains a SPARC processor).

Such translations can be performed either on compiled time, as in the case of static compiled simulations, where the overhead is completely eliminated; or on load time, as in the case of dynamic compiled simulations, where the overhead is amortized over/by the loops which repeatedly execute the same code.

The compiled-based ISSs generate very acceptable performance levels pertaining to simulation speeds. However, they also have some drawbacks: they are not portable; they require a large capacity memory; and they create certain difficulties when modeling certain aspects specific to the target processor (e.g. interruptions).

Consequently, despite their productivity problems, the interpretation-based simulators are widely used to validate cycle-accurate hardware/software.

Native simulation. Native simulation is performed directly using the host machine without modeling the architecture of the target processor which is supposed to execute the software once the embedded system is finally implemented.

Since this technique uses a native compiler (i.e. the compiler of the host machine), the code which is specific to the target processor (e.g. the assembly code) cannot be executed. Therefore, the code to be validated must be written in a high-level language, which is independent from the target processor (e.g. C or C++ language). Consequently, native simulations of complex applications (including several tasks and an OS) require a simulation model for the OS. This is mainly due to the fact that the final implementation of an OS includes code which is specific to the target processor.

The main advantage of this technique pertains to the speed of the simulation. The main challenge when using this technique consists of defining the OS simulation model which may have an important impact on the accuracy of the simulation.

2.2 Simulation Models for Operating Systems

Recently, OS simulation models have become increasingly popular for rapid hardware/software validations. This section presents the main characteristics of current OS simulation models.

Abstract operating systems. In the case of abstract OS simulation models, the embedded OS is abstracted and it is provided by a simulator. The application uses a high-level language to perform a set of parallel tasks. Communication between the different tasks is guaranteed by the communication primitives provided by the language specifications and the scheduling is ensured by the simulator kernel.

Virtual operating systems. The virtual OS is a model that mimics the functionalities of the final OS. The objective is to rapidly execute the OS and the running application. The virtual OS is also abstracted. The main difference from the previous model is that, in this case, the functionalities of the final OS are respected.

Generally, the virtual OS design is independent from the code that implements the final OS.

Final implementation of the OS. Finally, complete and accurate simulations require the execution of the final OS. Not only does this enable the validation of OS functionalities, but it also makes it possible to debug the final code.

2.3 State of the Art for OS Validation

This section presents previous studies pertaining to OS simulations using either the ISS-based simulations or native simulations. In the case of native simulations, both abstract and virtual OS models are used. For the ISS-based simulations, the final implementation of the OS is used.

Native simulations using abstract operating systems. Abstract operating systems are generally provided by simulation environments (e.g. SystemC [Sys05] or SDL [Sdl87]). These environments permit rapid validations of the application functionalities. However, designers may lack accurate timing information and the embedded OS is not taken into account.

Native simulations using virtual operating systems. Currently, several approaches propose native simulations of embedded software using virtual operating systems.

CarbonKernel [Car05] provides a developmental tool based on a virtual, real-time OS model. SoCOS [Des00] enables the modeling of virtual operating systems. These approaches present a code equivalency problem: the code of the final OS implementation differs from the code of the virtual operating model (e.g. the task-scheduler code differs from the virtual model the final implementation). This triggers the need for a supplementary step to validate the final code of the OS. Virtual OSs are also hindered by portability problems. This makes it difficult to validate several implementations for the same OS. To simulate a particular implementation, designers must "customize" the virtual OS by adding new functionalities. Generally, customization is carried out manually, a time-consuming task that can be the source of various errors.

Wind River Systems Inc. produced VxSim, a simulation model for VxWorks [Win05]. This model is very useful for fast simulations of software applications. Unfortunately, it is not possible to model interactions between the target processor and other hardware components.

ISS-based simulation to validate final OS implementations. ISS-based simulations are indispensable to execute the OS final code as it permits the accurate validation of software components (including both the OS and the application). However, as shown in Section 2.1, this technique is not practical when validating complex software. Consequently, native simulations currently represent an interesting alternative for rapid validation of complex software components. The main challenge consists of providing accurate OS simulation models and coping with current problems of code equivalences, time estimations and modeling of hardware/software interactions.

3. The Organization of Operating Systems

Generally, an OS is organized as a set of service layers (where a service represents an elementary functional unit): application programming interface (API) layer, basic service layer, and hardware abstraction layer (HAL). For each service, several implementations are possible.

The various services provided by the OS may be either processor-dependent or processor-independent. The realization of a processor-dependent service is different for each processor (e.g. the implementation is given in an assembly code). The realization of a processor-independent service is identical for all processors.

Generally, most services are processor-independent: these services can represent up to 90% of an OS implementation.

Figure 5.1 illustrates the OS organization involved in this approach:

- *The API* layer includes the high-level system calls invoked by the software tasks.
- *The Kernel* layer repents the core of the OS. The kernel provides the basic functionalities which are required to run both the user applications and system applications. It is also in charge of managing hardware resources.

  The kernel includes the following sublayers of service:

  - *The boot* comprises the services required to boot the system: it initializes the processor registers, the interrupt vector array, the memory and stack addressing space. It also charges the memory kernel (it copies the kernel code in the mass storage to the RAM).
  - *The context switch* includes all of the services related to the context of different tasks (saving the state registers of the tasks to be suspended and restoring the state registers of the tasks ready to run).
  - *The scheduler* includes all of the services related to task scheduling. Several algorithms may be used for scheduling (e.g. priority-based algorithms, round robin algorithms).

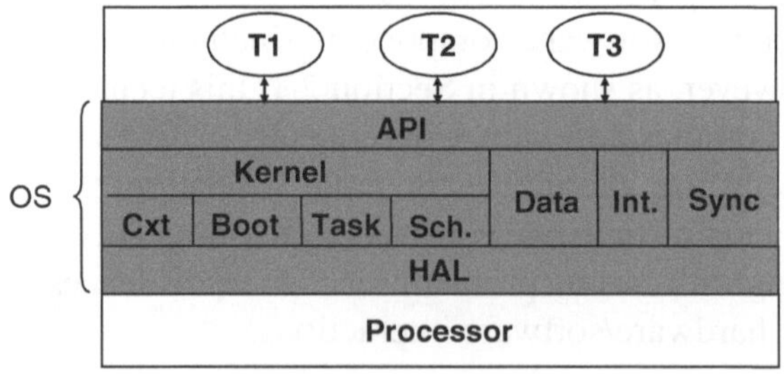

Figure 5.1. OS representation as a set of service layers

- *The task sublayer* consists of task management services. Generally, this sublayer establishes the link between the different sublayers of the kernel. It includes the task tables and the task structure.

  The kernel layer also includes a set of services that may be grouped into three main sublayers:

  - *The data sublayer* comprises all of the services required to manage the internal data structures (e.g. FIFOs) manipulated by the OS.
  - *The interruption sublayer* includes all of the services related to interruptions and system calls. These services respond to different hardware interruptions by initiating the right routines.
  - *The synchronization* sublayer includes all of the services related to the internal synchronization mechanisms of the OS.

- *The HAL* includes all of the services that allow communication between the OS and various hardware devices. It includes several groups of services, one of the most important being the services which provide device drivers for peripheral control.

4. Native Simulation Model for OS Final Code Validation

This section presents an OS simulation model for fast hardware/software validation. The objective is to provide a native simulation for software components while respecting the final implementation of the OS and providing temporal information.

4.1 OS Implementations vs. OS Simulation Models

The strategy used for the proposed simulation model is:

- To use the final code from the OS implementation and add temporal annotations to ensure accurate validation;
- To replace processor-dependent services with simulation models specifically designed for a simulation environment (e.g. SystemC, SpecC, C-Unix).

Figure 5.2 presents a comparison between the final implementation of the OS and its corresponding simulation model. Figure 5.2a shows the organization of the code for the final OS implementation. This particular example employs four services: Services 1 and 2 are implemented using a processor-independent code and Services 3 and 4 are implemented using a processor-dependent code. The simulation model corresponding to this OS implementation is illustrated in Figure 5.2b. In this model, the original implementation has been extended using

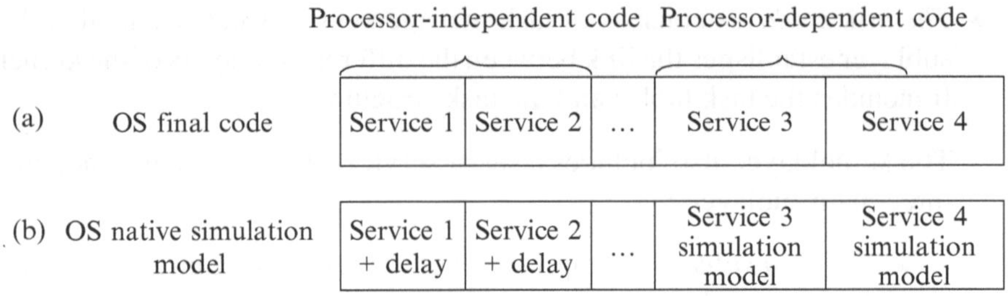

Figure 5.2. (a) OS implementation vs. (b) OS simulation model

timing annotations for Services 1 and 2. New functional models with timing annotations were designed for Services 3 and 4.

The utilization of the same code in the final implementation of the OS and in the simulation model considerably reduces the code equivalence problem presented in Section 2.3 for virtual operating systems. Thus, if the processor-dependent code represents 10% of the final OS implementation, 90% of the simulation model represents the final implementation.

4.2 Computing Execution Time Annotated in Simulation Models

To estimate the execution time required for timing annotations of he simulation model, conventional methodologies may be used [Ued02; Laj99; Suz96]. The execution time is estimated for the target processor and this value is subsequently inserted into the simulation model using a function that has the execution time as a parameter. This function, called delay, simulates the increment of a given time interval.

4.3 Simulation Models for the Target Processor's Independent Services

In order to obtain the simulation model for processor-independent codes, only the inserted timing annotations are required.

An example of a simulation model for a processor-independent service (the fifo_write communication service) is presented in Figure 5.3.

4.4 Simulation Models for the Target Processor's Specific Services

Some examples of processor-dependent services are boot services, context switches, interruption service routines and the peripheral drivers. Implementing a simulation model is required for each processor-dependent service.

Delay function specific to target processor

```
// Example: fifo service

Void OS::fifo_write(int f_id, int data) {

        disable_interrupt(); delay(10);

        if( fifo_full(f_id) == true ) {
                enable_interrupt(); delay(5);
                block(f_id); // task execution is suspended.
                disable_interrupt(); delay(10);
        }

        write(f_id, data);
        enable_interrupt(); delay(5);
}
```

Figure 5.3. Processor-independent code in OS simulation model

```
__cxt_switch            ; r0, old stack pointer, r1, new stack pointer,
STMIA  r0!,{r0-r14}     ; saving registers of the current task
LDMIA  r1!,{r0-r14}     ; restoring the registers of the next task
SUB    pc,lr,#0         ; return
END
```

Context switch: assembly code for ARM7 processor

```
void context_sw(int cur_task_id, int new_task_id)
{
        delay(34);
        wakeup_event[new_task_id].notify();
        wait(wakeup_event[cur_task_id]);
        delay(3);
}
```

Context switch: SystemC simulation model

Figure 5.4. Context switch: assembly code vs. simulation model

Figure 5.4 presents the ARM7 assembly code for the context switch service and its corresponding high-level simulation model implemented using SystemC. In this model, a list of events (wakeup_event_list) is used for task suspension (using the wait() function) and task notification (using the *notify()* function). The required execution time is simulated using the *delay()* function.

Moreover, a high-level simulation model corresponds to the boot service. When the native simulation is started, for task serialization, each task is suspended while waiting for a synchronization event from the kernel (e.g. wakeup_event). Since the processor architecture is not simulated in the native mode, the various tasks' states (e.g. registers and stack) are not simulated either.

The proposed simulation model presents a code equivalence problem (presented in Section 2.3) for the processor-dependent simulation services. However, since the assembly code represents a small part of the OS, this problem is considerably reduced when the final implementation code for processor-independent services is reused.

4.5 Simulating Interruption of Service Routines

Simulation models for interruption of service routines represent an important component of the OS model. Implementing these elements has an important impact on the performance and the accuracy of the model.

Although each processor manages different types of interruptions, only certain segments of these interruptions are necessary to validate the OS. For instance, interruptions specific to the ARM processor are: reset, nondefined instruction, software interrupt (SWI), data abort, prefetch abort, IRQ and FIQ [Jag96]. Interruptions related to the OS validation are: SWI, IFQ, and IRQ; their model is required to simulate the OS model.

Figure 5.5a illustrates an exception-handler model in an OS simulation, an example of the SWI code of the ARM7 processor. This figure shows a SWI routine in assembly code (_SWI_Routine) and a C code section to signal a generic SWI function called __trap_trap with SWI number 0 (defined by swi(0)). When the function indicated on line 17 is signalled, the processor-execution jumps to the vector table element of the SWI before the SWI handler recorded on line 2 is executed.

For fast OS simulations, the model of such an exception handler includes a minimal set of elements. In the case of an ARM processor, the minimal set of elements is composed of processor-mode registers (CPSR and SPSR's) that contain control bits such as interruption masks specific to each processor mode.

Figure 5.5 shows a model of an SWI routine for OS simulation and its use in C code to simulate the SWI's signal. Two functions, SWI_Enter and SWI_Return,

```
1.  // Assembly code for SWI routine
2.  _SWI_Routine
3.  STMIA  r13,{r0-r14}^  ; Push USER regisers
4.  MRS    r0,spsr        ; Get spsr
5.  STMDB  r13!,{r0,lr}   ; Push spsr and lr_svc
6.  LDR    r0,[lr,#-4]    ; Load swi instruction
7.  BIC    r0,r0,#0xff000000
8.  BL     __trap_trap
9.  LDMIA  r13!,{r0,lr}   ; Pop return address and spsr
10. MSR    spsr_cf,r0     ; Restore spsr for swi
11. LDMIA  r13,{r0-r14}^  ; Restore registers
12.                       ; and return to user mode
13. NOP                   ; NOP
14. MOVS   pc,lr          ; Return from SWI
15. // C code to use SWI
16. __swi(0) void __trap_trap(int, int, int);
17. __trap_trap(0, id, 0);
```

(a)

```
// Simulation Model
// SWI entrance
SWI_Enter() {
CPSR_save = CPSR;
SPSR_save = SPSR;
CPSR = SVC;
}
```

```
// Simulation Model
// return from SWI
SWI_Return() {
CPSR = CPSR_save;
SPSR = SPSR_save;
}
```

```
// Simulation Model
// C Code for SWI utilisation
SWI_Enter(); delay(24);
__trap_trap(0,id,0);
 SWI_Return(); delay(23);
```

(b)

Figure 5.5. Interrupt Service Routine: assembly code and corresponding C code

model the entry and return operations pending the SWI routines. For instance, the SWI_Enter function corresponds to lines 3 to 7 of the code section. In these functions, only the mode register change (CPSR and SPSR) is simulated. In the C code that signals the SWI, both functions are added before and after the SWI signal shown in Figure 5.5b.

Other exception handlers such as HW interruption managers are modeled in the same way.

In order to take hardware interruptions into account, the *delay()* function, used to simulate time increments, is enabled for this type of interruption. Thus, if a hardware component requests an interruption, the delay function takes this event into account and signals the simulation model to enact the interrupt service routine (ISR). To illustrate, Figure 5.6a shows an interruption which occurred during the execution of the *delay()* function simulating the increment of 10 clock cycles. The interruption occurs after the simulation of 5 clock cycles. As a consequence of this event, the *delay()* function signals the simulation model of the corresponding ISR. After the execution of the ISR, the global simulation time is advanced to the execution time required for the ISR execution (20 cycles in this example). Then, the *delay()* function simulates the value of the remaining time intervals (5 cycles in this example). If the *delay()* function execution is free from interruptions, the function simulates the value of the global simulation time with an interval given as the parameter (see Figure 5.6b).

The frequency of which the *delay()* function is signalled can determine the timing accuracy of simulated hardware interruption handling. However, the excessive execution of this function can also hinder the performance of the simulation. Thus, in the simulation flow presented, the designer can locate *delay()* functions by trading off simulation performance for accuracy. In terms of modeling task preemptions, [Lah00] presents a method to model the handling

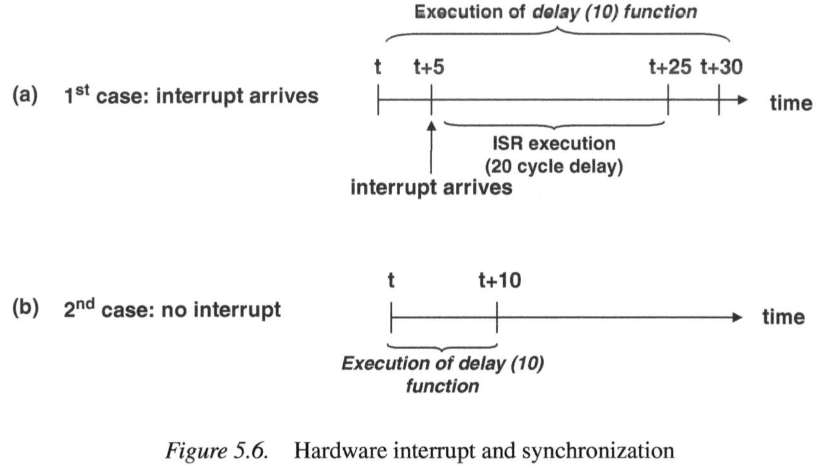

Figure 5.6. Hardware interrupt and synchronization

of interruptions. In this study, processor modes were not modeled separately and the author assumes that the order of the task execution does not change due to the handling of interruptions. In our modeling method, ISRs can signal task schedulers to invoke new tasks before returning to the task execution preempted by the interruption.

5. Automatic Generation of OS Simulation Models

Hardware/software interface designers have already proposed various methodologies to generate and/or configure an automatic OS. Thus, the automatic generation of OS simulation models may be achieved by reusing previous experiments in this field.

This section addresses the extension of an existing tool that enables the automatic generation, for a final implementation, of the OS by adding features required for an automatic generation of the OS simulation models. The original OS generator was presented in [Gau01]. The next section briefly introduces this tool.

5.1 Automatic Generation of Operating Systems

The main purpose of the methodology presented in [Gau01] is to find the required OS services for a given application in order to generate its code according to the target processor and, finally, to obtain a customized OS by composing the generated services.

Figure 5.7 illustrates, using an example, how the different services required by an application may be found. In this figure, the oval designs represent services (and their codes) and the rectangles refer to code sections associated with these services. For instance, the Priority, Load, Wait, Wakeup services use the Scheduling code section and the Context Switch service. Such a

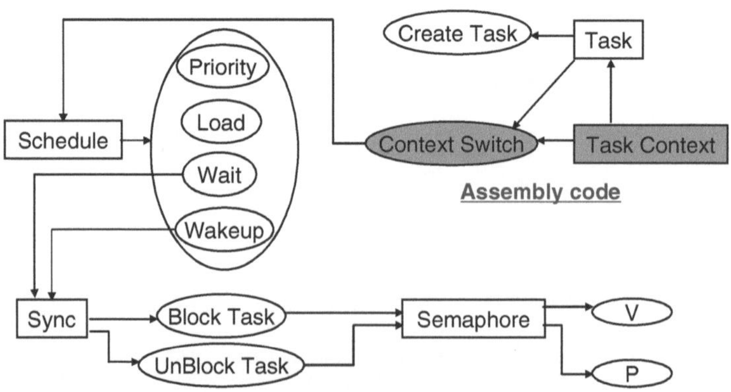

Figure 5.7. Service interdependency example

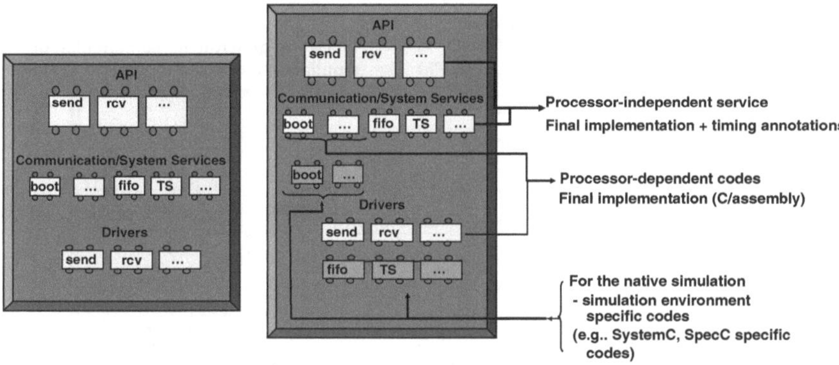

Figure 5.8. (a) Library to automatically generate OS; (b) Library to automatically generate simulation models

relationship may be transitive. For instance, semaphore services P and V use the four scheduling services (Priority, Load, Wait, Wakeup) through BlockTask and UnblockTask services and code sections Sync and Semaphore. Thus, according to the dependency chain illustrated, if semaphore services P and V are used by the application, all service codes and related code sections shown in the figure must be included in the OS in order to be generated. The code generated may either be processor-dependent or processor-independent.

Operating systems are generated using a library. This library (see Figure 5.8a) respects the organization of the OS, as it was presented in Section 3.

5.2 From Automatic Generation of OS to Automatic Generation of OS Simulation Models

The methodology presented in the previous paragraph may be adapted for the automatic generation of OS simulation models. This requires the extension of the library by adding elements for OS simulation models. Also, the addition of timing annotations is necessary for processor-independent services and the addition of simulation models is required for processor-dependent services.

Figure 5.8b shows the modified OS generation library used to generate simulation models.

Figure 5.9 illustrates the integration of the OS simulation model in the design flow of the hardware/software embedded systems. The OS simulation model is used for the rapid validation of the automatically generated OS. Only a small part of the OS will be validated using the ISSs. This accelerates design processes for the hardware/software systems.

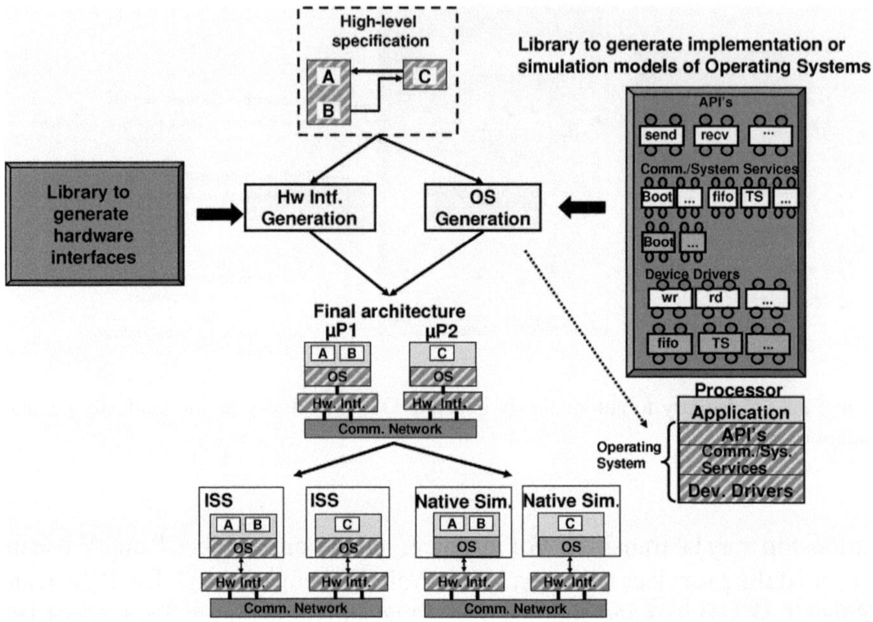

Figure 5.9. Integrating the OS simulation model into the design flow of embedded systems

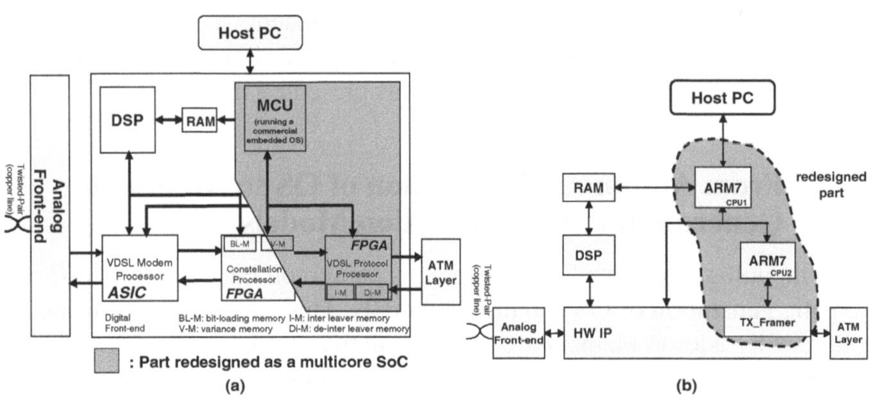

Figure 5.10. VDSL modem applications

6. Application of the OS Simulation Model

The OS simulation model presented was applied to the design of a VDSL modem as shown in Figure 5.10. Part of the system was designed using two ARM7 processors. The part designed as a multiprocessor system is shown on the left (see Figure 5.10), inside the grey box. The VDSL core functions, the analog interface and the DSP core are implemented in a third-party block. The DSP and the ASIC block execute functions such as (I)FFT, Reed-Solomon (de)coding, (de)scrambling, and (de)framing. In order to configure, monitor, and synchronize the DSP and the ASIC block, the control tasks, the host interface tasks, and the high-level VDSL code is mapped onto two ARM7 processors (CPU1 and CPU2 on Figure 5.10). CPU1 runs three concurrent software tasks while CPU2 runs six concurrent software tasks. The generating tool presented above was used to generate the OS. SystemC was the selected simulation environment. A software task was mapped to a thread in SystemC. Since SystemC provides dynamic sensitivity in the module member functions, an RPC process was mapped to a member function of the OS model in SystemC.

Two types of simulations were ran: one using two ISS's (one for each ARM processor) and the other using the generated OS models. For the remaining hardware components, the same simulation models in SystemC were used.

In these experiments, the generated OS simulation models gave more than two orders of magnitude simulation speed-up compared to the use of ISS's. When the number of ISS's is superior to two, this speed-up will increase even more due to the synchronization overhead between multiple ISS's.

7. Conclusion

Software components of embedded systems have become increasingly complex and the use of operating systems is mandatory to deal with concurrent and inter-task communication. This evolution makes hardware/software validation difficult while the performance of classical techniques (ISS-based) become more and more problematic. In this context, the definition of new validation techniques based on native simulation becomes mandatory. The key issue for this type of validation is the definition of OS simulation models required to enable fast and accurate simulation. Current OS simulation models present code equivalency problems or do not provide timing information and the possibility to validate hardware/software.

This chapter introduced the main techniques for software validation and existing approaches for OS modeling. A new OS simulation model was also presented. This model enables native simulation of software components while respecting the final implementation of the OS as much as possible. The proposed model allows for timed simulation and hardware/software validation. An existing tool to automatically generate operating systems was extended in order to automatically provide the simulation models for custom operating models. The simulation speed-up noted by the experimental results shows the effectiveness of the simulation model.

References

[Car05] Carbonkernel, available at http://www.carbonkernel.org.

[Des00] D. Desmet, D. Verkest, H. de Man, "Operating System Based Software Generation for Systems-on-Chip", Proc. Design Automation Conf. (DAC), June 2000.

[Gau01] L. Gauthier, S. Yoo, A. A. Jerraya, "Automatic Generation and Targeting of Application Specific Operating Systems and Embedded Systems Software", Proc. Design Automation and Test in Europe, 2001.

[Jag96] D. Jaggar, *Advanced RISC Machines Architectural Reference Manual*, Prentice-Hall, 1996.

[Lah00] K. Lahiri, A. Raghunantan, S. Dey, "Efficient Exploration of the SoC Communication Architecture Design Space", Proc. International Conference on Computer Aided Design, pp. 424-430, Nov. 2000.

[Laj99] M. Lajolo, M. Lazarescu, A. Sangivanni-Vicentelli, "A Compilation-Based Software Estimation Scheme for Hardware/Software Co-Simulation", International Workshop on Hardware Software Codesign, 1999.

[Row94] J. Rowson, "Hardware/Software Cosimulation", Proc. Design Automation Conference (DAC), 1994.

[Sdl87] Computer Networks and ISDN Systems. CCITT SDL, 1987.

[Suz96] K. Suzuki, A. Sangivanni-Vicentelli, "Efficient Software Performance Estimation Methods for Hardware/Software Design", Proc. Design Automation Conference, June 1996.

[Sys05] D. C. Black, J. Donovan, *SystemC: From the Ground Up*, Springer, 2004.

[Ued02] K. Ueda, K. Sakanushi, Y. Takeuchi, M. Imai, "Architecture-level performance estimation method based on system-level profiling", Proc. Design Automation and Test in Europe 2004.

[Win05] Wind River Systems Inc., "VxWorks 5.4", available at http://www.wrs.com/products/html/vxwks5.4.html/.

[Zhu99] J. Zhu, D. D. Gajski, "A Retargetable, Ultra-Fast Instruction Set Simulator", DATE Conference 99, Munich, Germany.

[Ziv96] V. Zivojnovic, H. Meyr, "Compiled Hw/Sw Cosimulation", Proc. Design Automation Conference (DAC), 1996.

Chapter 6

Anatomy of a Continuous/Discrete System Execution Model for Timed Execution of Heterogeneous Systems

F. Bouchhima,[1] L. Gheorghe,[1] M. Abid,[2] G. Nicolescu[1]
[1]*Ecole Polytechnique de Montreal,* [2]*ENIS, Tunis*

1. Introduction

Modern embedded systems like micro-electro-mechanical systems (MEMS), mixed-signal systems and real-time controllers integrate discrete and continuous components. These systems can be found in various domains such as health, communication, as well as the defence, and automotive industries. The design process of these systems needs to be improved, International Technology Roadmap for Semiconductor (ITRS) announcing a "*shortage of design skills and productivity* arising from a lack of training and poor automation with a need for basic design tools" as one of the most daunting challenges in the domain [ITR06].

Designers currently build the different discrete and continuous components by using powerful existing tools specialized for an application domain (e.g. SystemC or VHDL for the electronic digital part which is discrete, Matlab/Simulink for the mechanical part which is continuous) and they are generally not too fond of changing their tools. New CAD tools, enabling the global execution of continuous/discrete systems, are consequently mandatory. These tools must be based on global execution models which are independent from the specification languages and simulators and provide several implementations enabling integration of different existing tools in order to exploit their features.

One of the main difficulties in the definition of new CAD tools for continuous/discrete (C/D) systems pertains to the heterogeneity of the concepts manipulated by the discrete and continuous components. In the case of validation tools, the key issue consists of defining global execution models, accommodating several of the discrete and continuous execution semanticists. Co-simulation

G. Nicolescu and A.A. Jerraya (Eds.), Global Specification and Validation of Embedded Systems – Integrating Heterogeneous Components, 75–108.

(as defined in Chapter 4) is currently the most popular validation technique for heterogeneous systems. This technique was successfully applied for hardware/software discrete systems [Val95], but rarely applied to C/D systems. It allows a joint simulation of heterogeneous components with different simulation models.

This chapter proposes the anatomy of a C/D global execution. This model is based on a generic architecture integrating several synchronization models, which provide adequate accuracy/performance compromises. These models result from a discrete and continuous model study and the concepts involved in continuous/discrete simulation such as time and event management.

The remainder of this chapter is organized as follows: the following section introduces continuous and discrete simulations as well as the main concepts pertaining to such models. Section 3 addresses time distribution in timed simulation models and defines a time model for the global execution model. Section 4 discusses events management in continuous/discrete systems and Section 5 defines new synchronization models and a generic simulation model for continuous/discrete systems execution. Section 6 proposes implementation solutions for the defined execution model. Section 7 presents a tool for continuous/discrete execution. The application of the execution model for the validation of three heterogeneous systems is discussed in Section 8. Section 9 concludes the chapter.

2. Continuous Simulation Model vs. Discrete Simulation Model

Discrete and continuous systems are characterized by different physical properties and modeling paradigms. For example, mechanical systems and analog circuits respectively, are usually modeled by analog equations derived from Newton's and Kirchhoff's laws while discrete system models are based on mathematical logic resulting from Boolean logic and arithmetic expressions.

2.1 Discrete Systems Modeling and Simulation

This section introduces the concurrent processes formalism used in discrete systems modeling and presents their execution model.

Discrete systems are commonly modeled by concurrent processes describing their behavior using Boolean and/or algebraic expressions. These processes are generally grouped into modules according to their functionalities within the system. Modules are connected by signals through input/output ports.

Discrete model execution is based on *events*. An event represents an occurrence or happing of significance to a process. The process may wait for an event or any set of events or it may (request to) receive asynchronous notification that an event has occurred. For instance, modifying a signal value at a given moment causes an event. In this case, the event is represented by a couple (signal value, time of occurrence). Events represented only by their time of occurrence (e.g. clock event) are called *pure events*.

Processes are considered event-sensitive when events trigger executions. If several processes are sensitive to one or several events (with the same time of occurrence) then, these processes must be executed in parallel. However, executions often occur on sequential machines which can only execute a single instruction at a time, therefore one process.

Thus, this type of execution cannot really parallelize processes. The solution consists of emulating parallelism, which is based on a simple yet effective strategy: in order to execute each process "as if" the parallelism were real, it is necessary that its environment (its inputs) does not change when executing other processes. Thus, the process execution order loses its importance and everything takes place as if a parallel execution occurred. This requires that shared variables (signals) between processes keep their values until the execution of all parallel processes finishes.

Processes preserve the sequential aspects. The same instructions and control structures are found with C, Pascal and Ada languages. The only exception concerns the signal assignment: expressions assigning variables to signals are considered parallel expressions. Each assignment expression can thus be seen as a process. An example of two processes with Signals A, B, and C appears below.

Process 1
A <= B **and** C;
B <= Init **when** Select = '1' **else** C;
C <= A **and** B;

Is equivalent to:

Process 2
C <= A **and** B;
A <= B **and** C;
B <= Init **when** Select = '1' **else** C;

During simulation, the simulator must maintain *a timer* and associate a notification time for each event. Its main role consists of maintaining the event order in a global queue according to their notification times. A simulation cycle is performed at each discrete time. Within a simulation cycle, the event with the first ordered time stamp in the event queue is processed and the processes sensitive to this event are executed. This may generate other events that trigger the execution of other processes. Once all events with a time stamp identical to the current time are treated, the simulator advances its time to the value associated to the nearest discrete scheduled event and starts a new simulation cycle.

Process execution does not advance the simulator local time. Consequently, within the same simulation cycle the "cause" and the "effect" events will share the same time of occurrence, which violates the causality principle. To address this problem, the simulator uses a virtual time interval, called *delta*, the duration of which equals zero. The role of a delta-cycle is to order "simultaneous" events within a simulation cycle, i.e. identifying which event caused another. Thus, event "causes" and "effects" are consistently differed by a delta. Simulation cycles are composed of several deltas. The simulator uses a delta counter set at zero before launching a simulation cycle. If the processes executed at the beginning of the simulation cycle generate events, then the simulator annotates these events with a time stamp indicating the "real" current time with an additional delta. If there are no processes to execute at the current time ("real" current time plus zero delta), the simulator increments the delta counter (current time equals the "real" current time plus one delta) and executes the processes which are sensitive to the indicated events.

Although discrete event simulators can virtually execute any discrete system, their concept of rigorous event order may not be necessary. In fact, simulator variants have been proposed to increase discrete system performance:

- Synchronous systems simulation does not require global event sorting as signals can process events only as the clock ticks. Within a clock cycle, events can be totally or partially ordered, or entirely unordered, depending on the model [Cha96].

- Data flow model simulation solely uses partial ordering of events. Events associated to the different signals may not have ordering relationships. The advantage of this method: it avoids over-specifying a design when complete orders are not required [Cha96]. In [Pat04], this solution makes it possible to improve a discrete event scheduler by increasing synchronous data flow simulation performance.

2.2 Continuous Systems Modeling and Simulation

This section introduces the computation models used for continuous systems modeling. Their simulation model is also presented.

Modeling continuous systems. A continuous system is generally described using ordinary differential equations (ODEs). An ODE refers to an equality involving a function and its derivatives. ODEs can be written in the form (1) as a system of first-order ordinary differential equations. Higher order ODEs can be reduced to a system of first-order equations. Although higher order equations solved directly are sometimes more efficient, few tools are yet available for this purpose [Gup85].

$$\dot{y} = \frac{dy}{dx} = f(x, y), \quad y(x_0) = y_0 \qquad \text{where } y \text{ is a vector} \tag{1}$$

The form (1) is called explicit ODE. Another ODE form is the fully implicit ODE:

$$f(x, y, \dot{y}) = 0 \tag{2}$$

Most fully implicit ODEs can be written as [Gup85]:

$$M(x, y)\, \dot{y} = f(x, y) \quad \text{where } M \text{ is a matrix.} \tag{3}$$

The form (3) is called linearly implicit ODE. The inversion of the matrix M converts this form to the conventional form (1).

In the case of continuous systems, the form (1) becomes:

$$\dot{x} = \frac{dx}{dt} = f(x, u, t), \quad x(t_0) = x_0 \tag{4.1}$$

$$y = g(x, u, t) \tag{4.2}$$

where, t is the time, u is the inputs vector, x is the state variables vector, and y is the output vector.

Thus, a *state space* completely specified by the equations (4.1) and (4.2) is obtained. Formula 4.1 gives the set of *state equations* with its initial conditions and 4.2 indicates the set of *output equations*. Assuming there are n state variables, m input variables, and r output variables, these equations can be written in scalar form as a set of n state equations and r output equations:

$$n \text{ state equations} \begin{cases} \dot{x}_1 = f_1(x_1(t), \ldots\ldots, x_n(t),\ u_1(t), \ldots\ldots,\ u_m(t),\ t),\ x_1(t_0) = x_{10} \\ . \\ . \\ \dot{x}_n = f_n(x_1(t), \ldots\ldots\ldots, x_n(t),\ u_1(t), \ldots\ldots,\ u_m(t),\ t),\ x_n(t_0) = x_{n0} \end{cases}$$

$$r \text{ output equations} \begin{cases} y_1 = g_1(x_1(t), \ldots\ldots, x_n(t),\ u_1(t), \ldots\ldots,\ u_m(t),\ t) \\ . \\ . \\ y_r = g_r(x_1(t), \ldots\ldots, x_n(t),\ u_1(t), \ldots\ldots,\ u_m(t),\ t) \end{cases}$$

Linearity. The nature of the functions f and g in equations (4.1) and (4.2) consist of classifying *linear* and *nonlinear* systems. The system is said to be linear if its functions are both linear. In this case, the model given by (4.1) and (4.2) is reduced to the following system:

$$\dot{x}(t) = A(t)\, x(t) + B(t)\, u(t) \tag{5.1}$$

$$y(t) = C(t)\, x(t) + D(t)\, u(t) \tag{5.2}$$

$A(t)$ (n, n), $B(t)$ (n, m), $C(t)$ (r, n) and $D(t)(r, m)$ are matrices with n, m and, r are specified above.

Time-Invariant Systems. For time-invariant systems, functions f and g do not explicitly depend on time. In this case, equations (4.1) and (4.2) become:

$$\dot{x}(t) = f(x(t),\ u(t)) \tag{6.1}$$

$$y(t) = g(x(t),\ u(t)) \tag{6.2}$$

Assuming this time-invariance property, for linear system whose matrices $A(t)$, $B(t)$, $C(t)$ and $D(t)$ are all constant, we obtain:

$$\dot{x} = A\,x + B\,u \tag{7.1}$$

$$y = C\,x + D\,u \tag{7.2}$$

Differential-Algebraic Equations. If the set of equations describing the continuous system consists of both algebraic and differential equations, the equations are called differential-algebraic equations (DAE). The equations may be written as

$$\begin{aligned} &F_1(x, y, z, \dot{y}) = 0, \quad y(x_0) = y_0 \\ &F_2(x, y, z) = 0, \end{aligned} \tag{8}$$

where F_1 is a set of N equations and F_2 of M equations, with $N \geq M$.

Example. Figure 6.1 illustrates a simple continuous system, an RLC circuit.

The second-order differential equation describing this circuit is given by equation (9):

$$V_{in} = LC\frac{d^2V_{out}}{dt} + RC\frac{dV_{out}}{dt} + Vout \tag{9}$$

To solve this equation numerically, it must be rewritten as an equivalent system of first-order equations:

Making the assumption expressed by equation (10) and combining it with equation (9), the following system of first order equations is obtained:

$$\begin{aligned} y_1 &= V_{out} \\ y_2 &= \dot{V}_{out} \end{aligned} \tag{10}$$

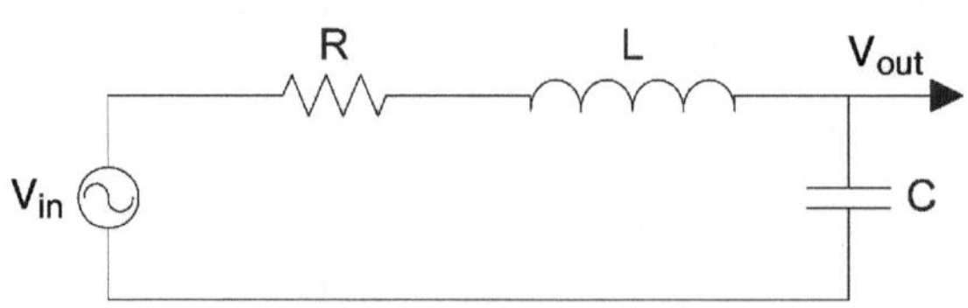

Figure 6.1. RLC circuit

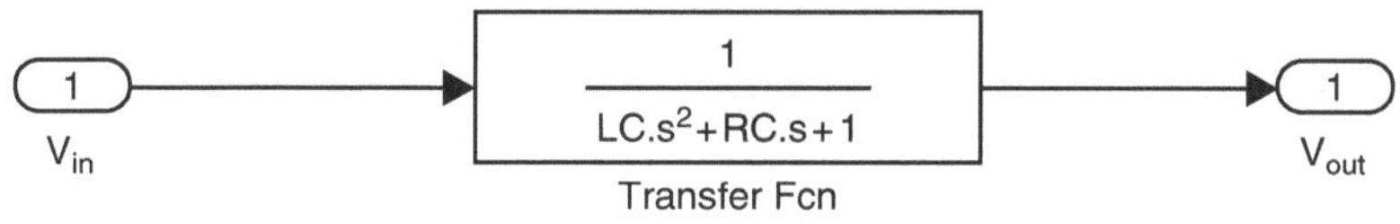

Figure 6.2. Modelling RLC circuit as a block, using the transfer function

$$\begin{cases} \dot{y}_1 = y_2 \\ \dot{y}_2 = 1/LC\,(V_{in} - RC\,y_2 - y_1) \\ V_{out} = y_1 \end{cases} \tag{11}$$

The same circuit can be easily described by the first-order ODE given by Forms 7.1 and 7.2:

$$\begin{bmatrix} \dot{x}_1 \\ \dot{x}_2 \end{bmatrix} = \begin{bmatrix} 0 & 1/C \\ -1/L & -R/L \end{bmatrix} \begin{bmatrix} x_1 \\ x_2 \end{bmatrix} + \begin{bmatrix} 0 \\ 1/L \end{bmatrix} V_{in} \tag{12}$$

$V_{out} = [1, 0] \begin{bmatrix} x_1 \\ x_2 \end{bmatrix}$, where $x_1 = V_c$ and $x_2 = I_l$ are the state variable

The system described by equation (11) and the system specified by equation (12) are equivalent. These systems can be modeled by using text editor or special blocks (the majority of continuous simulators provide special blocks).

Figure 6.2 shows the circuit from Figure 6.1 modeled as a single block specified using its transfer function given by the expression (13):

$$H(s) = \frac{V_{out}(s)}{V_{in}(s)} = \frac{1}{LCs^2 + RCs + 1} \tag{13}$$

Using (12) or (13), the circuit can be described using primitive blocks: the adder, the gain and the integrator blocks. The integrator represents the principal block [Cal91] (see Figure 6.3).

Continuous models simulation. Simulating a continuous model requires solving numerically differential and algebraic equations. A widely used class of algorithms discretizes the continuous time line into a set of ordered discrete time instants and computes numerical values of model variables at these ordered time instants. The interval between two consecutive time instants is called the *integration step* and, depending on the algorithm used, this step can be either fixed or variable. As explained in the previous section, continuous systems can be modeled using several blocks. During the simulation, the equations modeling the blocks composing the system are solved at each integration step, according to the order determined by the data dependence rule.

The criteria used to select integration steps are the accuracy, stability, and signal continuity. In the case of accuracy, both fixed and variable integration

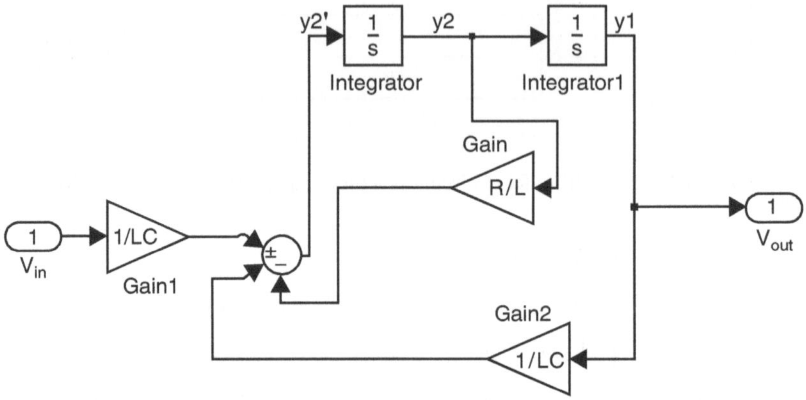

Figure 6.3. Modeling RLC circuit as a set of primitive blocks

steps can be used. However, the variable step algorithm improves the simulation performance. It reduces the step size and increases accuracy when model states change rapidly and it increases the step size to avoid unnecessary steps when the model states change slowly. When continuous models present discontinuities in form of finite jumps and/or stability problems, it is necessary to use:

- Variable step algorithms – to solve discontinuity problems observed at the solutions level, especially when interacting with discrete environments where signals change values discontinuously. In case of discontinuity, the algorithm reduces the step size again and this process can be repeated several times before the trouble spot is passed successfully [Gea84].
- Specific algorithms with a variable step – to resolve *stiff* [1] nonlinear systems of equations were used in the case of mechanical, thermal, and other models. This system exhibits time constants whose values differ by several orders of magnitude. Algorithms not designed for *stiff* problems are inefficient as they control the step size with accuracy rather than stability requirements [Cha96].

2.3 Heterogeneity in Continuous/Discrete Systems

Table 6.1 presents the main concepts characterizing continuous and discrete models: the notion of time, the means of communication and the process activation rules. Considering these concepts is a key issue for composing these two

[1] If, in a certain interval of integration, a numerical method is forced to use a step length which is excessively small in relation to the smoothness of the exact solution in that interval, then the problem is said to be *stiff* in that interval.

Table 6.1. Basic concepts for discrete and continous simulation models

Concepts / Model type	Time	Communication means	Process activation rules
Discrete	• Global notion for all system processes. • Advances discretely according to event time stamps.	Set of events located discretely on the time line	• Processes are sensitive to events. They can be executed in parallel. • At a given discrete time (simulation cycle), the execution order is determined using the delta concept.
Continuous	• Global variable involved in data computation. • Advances by integration steps.	Piecewise-continuous signals	• Processes (blocks) are executed sequentially at each integration step. • The order of execution is determined by using the technique known from static data flow (data dependency).

types of simulation models. Their composition must preserve the accurate data communication between them. This requires time synchronization and signal adaptation.

The next section introduces the time distribution approaches used to distribute time among processes. It also fixes the time distribution model for the global simulation model involving discrete/continuous simulation. This time model will be used when proposing the synchronization models.

3. Time Distribution Approaches

When designing hardware or software embedded systems, time is considered an important factor. In most cases, it must be considered with a high level of accuracy and measured with physical units. Timed simulation models assure that these requirements are fulfilled by providing a physical time model. Three approaches are proposed to calculate and distribute time among system processes [Jan04]: the local time counter approach, the time stamps approach and the absent event approach. An overview of these approaches appears below.

3.1 Local Timer Approach

This approach considers a local time counter for the entire model. In order to obtain the global time, each process accesses this counter. In a typical usage, a time-out period is set by the process and the time counter emits an event when the time period has expired. This approach is typically used by programming languages supporting real-time process modeling such as SDL-RT [SDL06], and others derived from C. Generally, the time is not defined by the language; it is supplied by a physical source of time, a timer or a quartz oscillator counter.

A correct functionality requires that all the processes share the same global time value at a given time instance. This is obvious for systems where all processes use the same clock (e.g. single processor systems). However, synchronization is required with distributed processes (e.g. multiprocessors) using various sources of time. Most hardware languages consider time as an integral part supplied by a virtual counter. A model composed of several processes described by different languages requires a synchronization mechanism for local time. Such synchronization is mandatory for accurate event exchanges between distributed processes. Figure 6.4 illustrates two processes P and P′ with their local times T and T′ respectively. If $T \neq T'$, then an incorrect event exchange will occur: event e is emitted by P at time T and consumed by P' at time $T' \neq T$ and vice versa.

3.2 Time Tag Approach

In this approach, there is no shared source of time. Each event exchanged between two processes is annotated with a time tag that represents the global time of the event occurrence. A process receives information on time only via explicit events on its input ports. It is clear that two processes will be synchronized if they receive two events characterized by the same time of occurrence. For instance, the two processes P and P′ illustrated in Figure 41 are synchronized if $t = t'$.

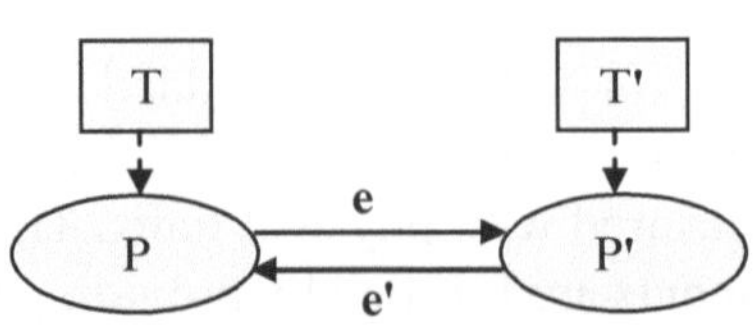

Figure 6.4. Distributed processes exchanging events

This approach increases process interdependency and eliminates unnecessary synchronization. However, it lacks total knowledge of the global time while not connected to a clock source process. Therefore, situations which require complete global time knowledge cannot be modeled with this approach.

3.3 The Absent Event Approach

In this approach, the concept of absent event enables time information communication. The time line is divided into discrete time slots and, at every time slot, all processes send events to all output signals. If no useful events are generated by a process at a given time instance, an absent event is emitted instead. The absent event value is distinct from all other possible values in the system. Its role is to inform the other processes about the time elapsed in the process that emitted it. Thus, all of the system processes are aware of the global time.

3.4 Summary

In conclusion, the first approach associated with synchronized processes is the most flexible. Although the second approach is interesting, as it preserves process independence properties, it remains applicable to a restricted number of models (used especially in data flow models). However, a time source could be added to this latter approach, but it would make it equivalent to the first approach. Finally, contrary to the third approach, where the time line is divided into discrete time slots, the first and the second approach can be used naturally for continuous time models.

4. Time Distribution Model Involved in Continuous/Discrete Execution Model

Continuous time and discrete execution models are timed models, where time is an explicit component integrated by the simulators. The continuous simulation model uses the local time counter-approach, while the discrete simulation model (discrete events) uses this approach with an additional time tag approach. At all simulation instants, both models can access the local time of the simulators on which they are executed. Thus, each process can annotate its emitted events with its own time stamps.

The global execution model cannot be characterized by a local time as this requires a tight synchronization between the continuous and discrete simulators. The approach based on event tags can be used for this type of global execution model. Thus, the events exchanged between the two simulators convey the time information. This avoids the need of a complete synchronization for each simulation time step. Using the event tag approach, a simulator performs synchronization only to accurately reach the time instants given by the event tags (stamps).

5. Global Execution of Heterogeneous Continuous/Discrete Systems

Continuous and discrete models interact via events. The instances of consuming and emitting these events must be considered with respect to causality and correct event exchanges. This section addresses event management in continuous/discrete systems co-simulation. We also propose a global simulation model based on accurate synchronization models.

5.1 Event Management in Continuous/Discrete Systems

In continuous/discrete systems, an event occurs when signal values change or once a variable exceeds a given threshold. In the first case, the event is defined by the couple (value, time stamp) while in the latter, it consists of a pure event solely defined by its time stamp. The events exchanged between the continuous and discrete models composing a continuous/discrete system are:

- The discrete model (DM) sends two types of events:
  - The *signals update events* caused by the modification of its discrete output signals
  - The *sampling events* which are pure events sent to the continuous model to indicate the sampling instants of its output continuous signals
- The continuous model (CM) can send *state events*. State events are pure and unpredictable events whose time stamps depend on the CM state variables (e.g. a zero-crossing event, a threshold overtaking event).

It is important to notice that the data path goes from the DM to the CM for signal update events, while it travels in the reverse direction (from CM to DM) for the case of sampling events.

Event exchanges must respect the causality principle during systems execution. A model is considered causal when causes precede effects. Here is one of the definitions for "causality": "The output of a process at time t should not depend on inputs that are later than t" [Liu02].

This section presents a study of event exchanges in the case of continuous/discrete global execution, where we examine compliance with causality and discuss incorrect situations that occur in event exchanges. Such situations are illustrated in Figure 6.5 and discussed below.

In Figure 6.5a, the DM generated an event at the time t_j while the CM was at the time t_i, where $t_i < t_j$. Since its time is continuous, the CM proceeds from this time and generates a state event at time t_k. This state event is caused by the consumed event at time t_j ($t_k < t_j$), violating the causality principle. In this case,

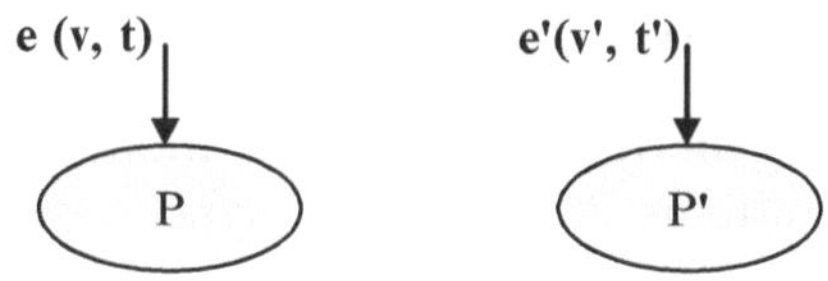

Figure 6.5. Processes synchronized by the time stamps approach

a possible solution consists of controlling the CM in order to take into account an event sent by the discrete model solely once its local time matches the time stamp of this event (each event is sent with a time stamp). Hence, even if a state event is generated at time t_k, the causality principle is not violated as this event was not caused by a discrete event not yet taken into account. However, this state event sent to the DM will have a time stamp that is inferior to the local time of the discrete simulator. This situation is illustrated on Figure 6.5c. In this case, in order to take into account the state event, the DM must be able to perform a rollback to a required time stamp. For most simulators, this rollback represents a daunting task since significant memory resources are required [Ree04]. To avoid the rollback, the CM local time must be consistently superior or equal to the discrete model local time. This solution is very effective. However, as further explained in Section 5.2, it may be preferable to run the DM before the CM for improved performance.

In Figure 6.5b, the CM generates a state event at time t_j. The DM is at the time t_i, where $t_i < t_j$. This case also presents a causality violation since the DM can generate an event at time t_k, where $t_k < t_j$. Note that the DM may generate an event regardless of the state event and this new event can impact the CM state variables and can subsequently cause a state event cancellation. The solution remains in controlling the DM to provide the time of each of its next output events. Then, the CM accurately reaches this time instant and sends its possible state events without any risks. This takes care of the causality problems.

Figure 6.5d illustrates another case where causality is violated. The CM is at time t_i and the DM generates an event at time $t_j < t_i$. In this case, an initial solution consists of using a continuous simulator that can perform rollbacks. However, this situation can be avoided when adopting the previously mentioned strategy: running the DM before the CM. Thus, the CM cannot exceed the time of the following input event from the DM. Indeed, it has to proceed until this time and then stop. Consequently, the DM must always provide the time stamp of its next event. This solution requires a DM able to predict the time stamp of its next output event.

This discussion shows that preserving causality is an important requirement to be considered when defining interfaces for the global execution model of a continuous/discrete model.

5.2 Synchronization Models

This section introduces a generic synchronization model for the global execution of continuous/discrete systems. An alternative synchronization model allowing the minimization of the interaction between the continuous and the discrete simulator is also presented.

Generic continuous/discrete synchronization model. Figure 6.6 gives an overview of the generic synchronization model, where the continuous simulator runs before the discrete simulator.

Assuming that the continuous and the discrete simulators are synchronized at the A time instant, the discrete simulator executes (without modifying time) all processes sensitive to the current notified events and updates signals (without modifying its time). Then it sends the time stamp of its next output event (point B, Figure 6.6) to the continuous simulator. Before increasing its time, it switches the simulation context towards the continuous simulator (arrow 1, Figure 6.6). The latter computes signals by solving differential equations until it accurately reaches the time sent by the discrete simulator (point C, Figure 6.6). Two cases are possible:

- This time corresponds to the occurrence time of a sample event. In this case, the continuous simulator updates the signals with the values calculated at this time and switches the simulation context to the discrete simulator (arrow 3). The latter will advance to the occurrence time of the stamp event (Figure 6.6, arrow 4) and restarts the cycle.
- This corresponds to the occurrence time of a signal update event. In this case, the continuous simulator stops temporally allowing the discrete

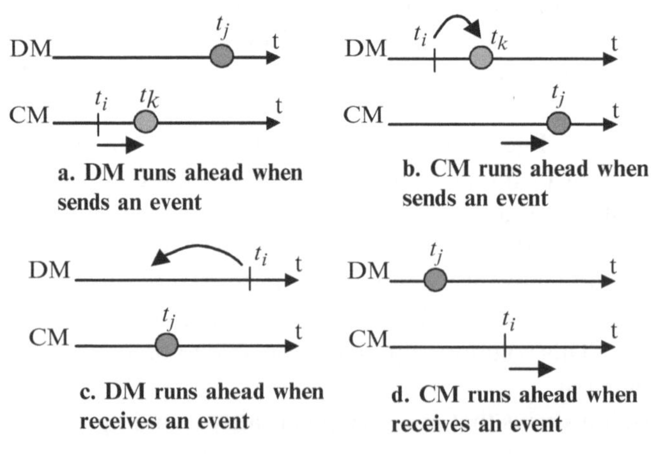

Figure 6.6. CM/DM events exchange

simulator to advance to the indicated event time stamp, to compute signals and send their values and the next event time stamp. Finally, it stops temporally allowing the continuous simulator to resume its execution. The latter proceeds to this time with the new signal values and the cycle starts over (Figure 43, arrows 5 and 6).

The continuous model may generate a state event. In this case, the continuous simulator indicates its presence, sends its time stamp to the discrete simulator and the simulation context is switched (Figure 6.6, arrow 7). The discrete simulator must be able to consider the state event by advancing the local time to the event time stamp and by executing the processes that are sensitive to it (since it is an external event).

One of the key features of this synchronization model is that it eliminates the need for rollbacks. As mentioned above, this requires that the discrete simulator can predict the time stamp of the next event corresponding to the following synchronization instant. Sampling events occur for each sampling period, so their time stamps are easily predictable. The difficulty is given by the signal update events.

Depending on the DM behavior, there are two modes which respect the presented synchronization model:

- The first mode, the *Predictable Events mode (PE)* can be used when signal update events are predictable (e.g. periodic events). In this mode, event time stamps and sampling events are placed and sorted in a special queue. In order to obtain the time stamp of the subsequent output event, the queue is consulted to find the minimum time stamp. Then, the type event (sampling or update signal) is verified and the information is sent to the continuous simulator.

- The second mode, the *full synchronization mode* (FS) can be used when signal update events are unpredictable. In this mode, the discrete simulator sends its next discrete time (always known) which may correspond to the time stamp of a signal update event. The synchronization overhead specific to this mode depends on the DM computation granularity.

Synchronization model for systems including unpredictable signal update events. In software models, signal computation requires several discrete steps. Generally, predicting the total number of steps is very difficult. In this case, the synchronization overhead of the FS mode may be inconvenient because the number of unnecessary synchronization steps will be much superior to the number of useful synchronization steps. Also, the PE mode cannot be used since signal update events are not periodic. The solution consists of running the discrete simulator in advance until it generates and sends an event to the

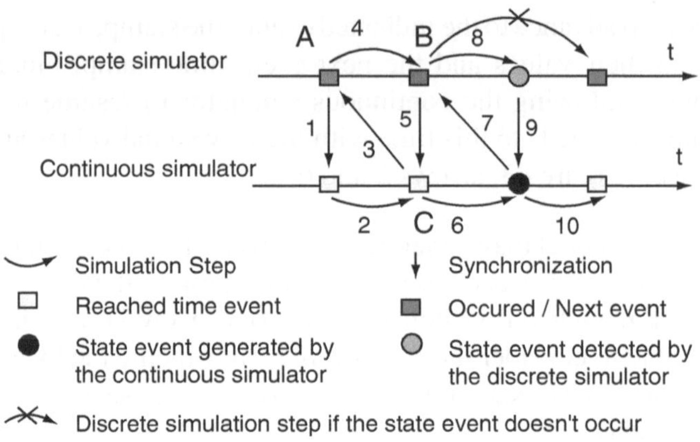

Figure 6.7. Generic continuous/discrete synchronization model

continuous simulator (with its time stamp). The synchronization model appears in Figure 6.7. This model defines a new mode which is the *unpredictable events mode* (UE).

Section 5.1 reported that, in this case, the causality problem can be solved by considering event time stamps. The difficulty is that the CM can generate a state event, requiring the discrete simulator to backtrack (Figure 6.6c). This model is recommended if the CM never generates state events as it eliminates all unnecessary synchronization. However, if the CM generates state events, rollbacks are necessary.

We suggest the use of this model, enhanced with state event consideration for control systems, where the DM represents a software component specified at the ISA abstraction level. State events model external interruptions. In this case, the checkpoint-based technique [Fle95] provides an effective solution, allowing light-rollbacks which require reduced memory resources. Indeed, only a backup of memory data segment, processor registers as well as input and output signal values will be made for each output discrete event time stamps used as checkpoints. In the case where the CM generates a state event, the discrete simulator performs a light-rollback toward the time stamp of the previous output event, restores the saved data, initializes the time counter with this time stamp and starts over with this time, taking into account the state events. The new counter replaces the discrete simulator local time and becomes the time source whose unit equals the period of the processor clock.

Summary. Table 6.2 shows the possible synchronization modes that can be used according to the continuous and the discrete model. The FS mode can usually be used but the overhead created by this mode may not be acceptable

Table 6.2. Synchronization modes depending on the continous and discrete models

Discrete Model \ Continuous Model	Predictable Events	Unpredictable Events
State Events	FS, PE	FS, UE with state event considerations (ISA level)
No State Events	FS, UE	

(e.g. at the ISA level). If the CM does not generate state events, then the UE mode is recommended since it avoids unnecessary synchronization.

5.3 Global Execution Model

To run continuous and discrete simulators with respect to the presented synchronization models, new simulation interfaces must be added to allow detecting:

- State events generated by the continuous model and the consideration of these events by the discrete simulator
- Discrete simulator events (by the continuous simulator)
- The end of the discrete simulation cycle and event sending

A generic architecture for the global continuous/discrete execution model is illustrated in Figure 6.8. Continuous and discrete execution models communicate through a *co-simulation bus* via *simulation interfaces* (as introduced in Chapter 4). For both models, these interfaces implement two main *layers*: *the synchronization layer* and *the communication layer.*

Communication is assured by two layers: “data exchange” and “signal conversion and data exchange”. Data exchange consists of reading or writing signal values that connect both models. Signal conversion consists of converting continuous signals to discrete signals and vice versa. The synchronization layer will be detailed in the following section.

5.4 The Synchronization Layer

This layer consists of six sublayers (Figure 6.8). Their task consists of supplying mechanisms that can address the three difficulties presented in Section 5.3. Their role and response to each difficulty is explained in the following sections.

State event detection and consideration. Most continuous simulators provide adequate mechanisms to detect the state events generated by the continuous model. Once detected, this event is sent by the “detection and sending of state events” layer. The discrete model “State events consideration” layer must be

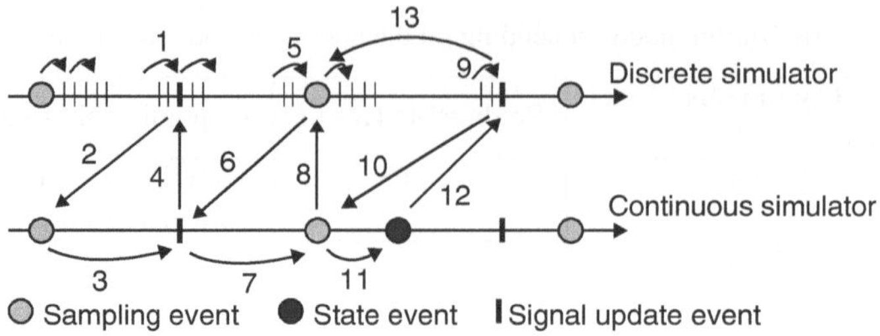

Figure 6.8. Handling of unpredictable updates

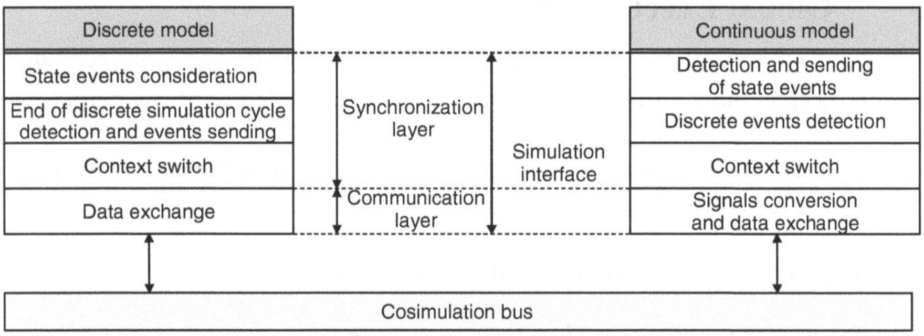

Figure 6.9. Global Execution for accurate continuous-discrete simulation model

able to take this into account. Then, the next discrete time must reflect the event time stamp rather than the internal event time stamp scheduled by the discrete simulator (Figure 6.6). In the case of the UE mode (Figure 6.7), this synchronization layer must initialize the new timer (see Section 6.5.2.2) with the last sampling event time stamp (light-rollback) and to restore the saved data.

Detecting events from discrete model. The continuous simulator must move forward until the discrete model event is detected (without missing it). This may not correspond with its discretization time. The "discrete event detection" layer (Figure 6.9) must force the continuous simulator, to adjust its integration steps which detect the event and satisfy the resolution criteria (accuracy, continuity, and stability), when approaching the event time stamp coming from the discrete model Figure 6.10 illustrates this phenomenon. It is also possible to use a fixed integration step that can be changed to a variable step when the simulator comes near a discrete event time stamp.

Detecting the end of discrete simulation cycle. Most discrete events simulators (such as the VHDL [IEE99] and SystemC [Sys03]) use the delta concept.

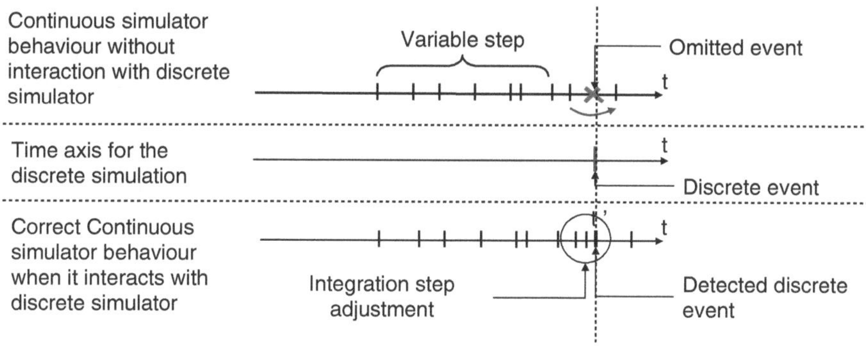

Figure 6.10. Detecting events from the discrete model

The problem of detecting the end of the discrete simulation cycles appears for this type of simulator.

For these simulators, delta-cycles are performed during delta time spans. They essentially contain two phases: an evaluation phase to execute processes and an update phase to update signals that were modified at the evaluation step. At a given discrete time, an unpredictable number of delta cycles may occur (Figure 6.11; Section 2.1) until the simulated model stabilizes: no signals to change, or in a general way, no more zero-delayed events to consider at the current time. Then, the discrete simulator increases its local time to the value of the following discrete time (next event time stamp). To guarantee that the context switch layer transfers the simulation control to the continuous model only once at a given discrete time, the discrete signals that connect both models have been stabilized, the "end of discrete simulation cycle detection and event sending" layer (Figure 6.9) becomes necessary.

6. Implementing the Global Execution Model

This section presents solutions to implement the synchronization and the communication layers for SystemC (used as an example of discrete event simulator) and Simulink (as an example of continuous simulator). To better illustrate, first, here is a brief presentation of SystemC and Simulink simulators.

6.1 SystemC

SystemC [Sys03] is a standardized modeling language intended to facilitate system level design and intellectual property exchange at multiple abstraction levels, for systems containing both software and hardware components. The SystemC simulator includes an effective and relatively simple scheduler. As indicated, the SystemC scheduler uses the delta-cycle concept (see Section 2.1). Its task consists of determining the order of the execution process by considering sensitivity lists and events in its global queue. The last one is ordered according to the time stamps of these events. The first element in the queue thus represents the next event to occur. Events are classified into two types: zero-delayed and timed events. The time stamp of a timed event translates into a next "real" time. The time stamp of a zero-delayed event consists of two components: the current "real" time plus the sum of the number of deltas: both components are used to order the events in the queue.

6.2 Matlab/Simulink

Highly popular and widely known by the modeling and simulation community, Simulink [Mat06] offers several libraries in the automotive and power electronics sectors, etc. and seven solvers designed for stiff (appearing in nonlinear systems) and nonstiff problems, which can provide the utmost accuracy. Simulink solvers subdivide the simulation time span into major and minor integration steps, where a minor integration step represents a subdivision of a major integration step. Solvers produce results for each major integration step, using the resolution results at the minor integration steps to improve result accuracy at major integration steps.

The order in which blocks are updated is critical for result validity. The data dependence rule is used during the initialization phase in order to statically determine the order of block activation. In fact, if block outputs are depend on its inputs, they must be updated after the blocks that drive their inputs (e.g. adder or gain computing block). This approach is called *direct-feedthrough.* All of the other blocks are called *nondirect-feedthrough* (e.g. integrator block). To assure a valid update order, Simulink uses the following rule: *nondirect-feedthrough* blocks can be executed, first in no particular order, followed by *direct-feedthrough* blocks in an order which respects the above-mentioned rule.

6.3 Implementing the Synchronization Layer

The following section details the implementation of the different synchronization sublayers in the global simulation model.

Implementing discrete event detection sublayer. Simulink does not make it possible to control integration variable steps in a direct manner. Consequently, it is difficult to guarantee accuracy to detect discrete events (the discrete event time stamp).

To cope with this difficulty, the "discrete event detection" sublayer was implemented in a special S-function. The last one is devoid of input or output ports. Its role is to create breakpoints that must be reached accurately by the solver (without going beyond). The time mode used in this S-function (VARIABLE_SAMPLE_TIME) allows for choosing its next time execution equal to

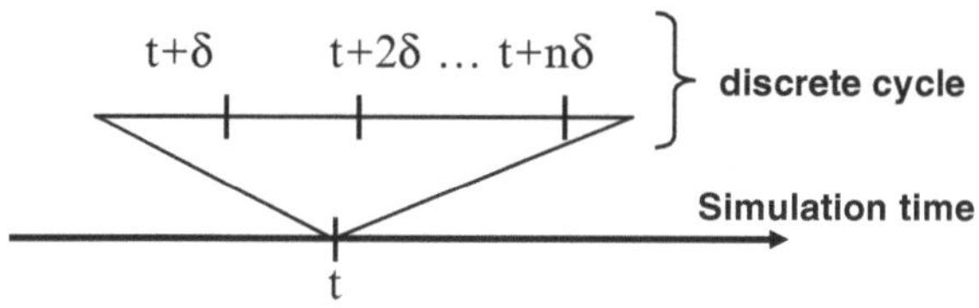

Figure 6.11. A simulation cycle is composed of an unpredictable number of delta cycles

```
/* Function: mdlGetTimeOfNextVarHit */
#define MDL_GET_TIME_OF_NEXT_VAR_HIT
static void mdlGetTimeOfNextVarHit(SimStruct *S)
 {

        /*time of the next breakpoint to consider by the solver */
        double NextBreakpointTime;

        /* read the sample event time stamp in memory*/
        double SamplingEventTime= *((double*)lpMapAddress + 200);

        /* read the signal update event time stamp in memory*/
        double UpdateEventTime =  *((double*)lpMapAddress + 100);
        if(SamplingEventTime < UpdateEventTime)
                NextBreakpointTime = UpdateEventTime;
        else NextBreakpointTime = SamplingEventTime;

        /* set the next breakpoint*/
        ssSetTNext(S, NextBreakPointTime);
 }
```

Figure 6.12. Function creating breakpoints

the next discrete event sent by the SystemC synchronization layer. In this case, Simulink adjusts the integration steps to satisfy the resolution criteria and to accurately reach the time execution of this S-function (which is the time stamp of the SystemC event). Once the sublayer detects that the event is reached, it sends (sampling event) or receives (signal update event) data. The code given in Figure 6.12 illustrates the function (member of the "discrete event detection" sublayer) that creates breakpoints.

Detection the end of discrete simulation cycle and events sending sublayers. To guarantee that SystemC sends data and transfers the simulation control to Simulink only after discrete signals have been stabilized, detecting the end of the discrete simulation cycle is necessary. Since SystemC does not provide such mechanism, a modification to its scheduler was made to detect the discrete simulation cycle end and to switch the simulation context to Simulink. This functionality was added to the *simulate()* function in the *sc_simcontext* class of the SystemC scheduler. This function essentially contains the simulation loop. Figure 6.12 shows the pseudo-code that provides a part of this function and indicates the end of the discrete cycle location.

Synchronization sublayers for state event cases. In the case of state events, there are two synchronization sublayers: the "state event consideration" sublayer (in the discrete model synchronization layer) and the "detection and sending of state event" sublayer (in the continuous model synchronization layer).

To detect state events, the "detection and sending of state events" layer (Simulink side) adds a "*Hit Crossing*" component from the Simulink library. This component compares the input signal to the hit crossing the offset value. If the signal increases, falls or remains at the offset value, the block output is set to "1". Once a state event is detected, the indicated layer communicates its presence by setting a special flag and sending its time stamp.

For the "state event consideration" layer (SystemC side), the solution consists of inserting a pure event (without value) into the SystemC simulator queue whose time stamp is equal to the time of the state event occurrence. The event insertion must occur before the discrete simulator increments its timer (Figure 6.7, between arrows 3 and 4). Otherwise, it must backtrack in order to take it into account. Figure 6.13 locates this insertion point of such an event. The scheduler modification solely consists of creating a set of events which can be notified in the case of state event presence. Their notification causes the execution of the SEC_Method in the code below.

For the "state event consideration" layer (the part implemented by the SystemC interface), the usual syntax to create processes sensitive to events was used (see the code below). From the designer point of view, a user process which is

1. *Initialization Phase* – Execute all processes (except SC_CTHREADs) in an unspecified order.
2. *Evaluate Phase* – Select a process that is ready to run and resume its execution.
3. If there are still processes ready to run, go to step 2.
4. *Update Phase* – Execute any pending calls to update() resulting from request_update() calls made in step 2.
5. If there are pending delayed notifications, determine which processes are ready to run due to the delayed notifications and go to step 2.
6. **If Mode = FS then Send the next discrete time to Simulink and Switch context to Simulink, else**

    If Mode = PE then send the next *signals update events* **or** *sampling events* **time stamp and Switch context to Simulink.**

    Else (mode = UE) if *signals update events* **flag = "1" then Switch context to Simulink**
7. **If state event, then add to the scheduler queue a timed event with time stamp equal to the state event time stamp.**
8. If there are no more timed notifications, the simulation is finished.

Figure 6.13. SystemC enhanced scheduler

```
#define et_mat0
sc_get_curr_simcontext()->et_mat[0]

/* this definition is in the file defining environment variables added for
heterogeneous simulation */
InterfaceIn.h
...
sc_out <sc_bit> StateEventPort;
SC_CTOR(InterfaceIn)
{
 ...
//creation of et_mat0 event associated with //state event
et_mat0 = new sc_event; // et_mat0 will be notified by SC scheduler

//make SEC_Method sensitive to et_mat0,
//as consequence to the state event
SC_METHOD(SEC_Method);
sensitive(et_mat0);

... }
InterfaceIn.cpp
......
Void InterfaceIn::SEC_Method()
{
StateEvPort.write(~StateEvPort.read());
}
```

Figure 6.14. "State event consideration" layer in SystemC interface

sensitive to a state event must be marked sensitive to the input signal (reserved for this state event) coming from the SystemC interface. The following code in Figure 6.14 illustrates an example of the "state event consideration" layer in the SystemC interface.

In the case of state events, Simulink indicates its presence to the SystemC scheduler, which notifies the event (here *et_mat [0]*) associated with it by the event time stamp. This notification causes the execution of the SEC_Method process (see Figure 6.14).

6.4 Implementing the Communication Layer

To ensure "data exchange" between simulators, a shared memory created by the file Mapping API from Windows was used. It has a defined structure composed of data ports which connect signals, flag ports and time ports to exchange event time stamps.

For "signal conversion", a sampler was used to adapt continuous signals to discrete components. A signal generator (zero-order hold) was also used to extrapolate events points in order to adapt discrete signals to continuous components. By nature, the shared memory can play the role of zero-order hold, consequently no component was added. The sampler consists of sampling signal values at synchronization points. In this case, its role is limited to writing signal values once the Simulink synchronization layer detects a sampling event from SystemC.

6.5 Discussion

The presented global simulation model remains independent from languages and environments. However, the modification of the discrete simulator was necessary during the implementation phase and discrete simulators must allow these modifications. If discrete simulators provide mechanisms to detect the end of the simulation cycle, modifications are unnecessary and commercial simulators can be used. For instance, in the case of the ModelSim VHDL simulator simulation interfaces are implemented as foreign-language interface (FLI) functions [Mod06]. In order to solve detection discrete cycles (i.e. switch simulation context only after discrete signals are stabilized), the "context switch" layer can be implemented by a VHDL process with a MTI_PROC_POSTPONED priority. Postponed processes (when triggered) run once at the end of the discrete cycle for which they are scheduled after all other processes. They can schedule an event in zero delay [Mod06].

7. CODIS a Co-Simulation Tool for Continuous/Discrete Systems

COntinuous DIscrete Simulation (CODIS) is a consistent tool which can automatically produce global simulation model instances for discrete/continuous systems simulation using SystemC and Simulink simulators. This is done by generating and providing interfaces which implement the proposed simulation model layers. It respects the presented synchronization models and offers various options for a most adequate mode. Figure 6.15 gives the overview of the flow of global simulation model generation. The inputs in the flow are the continuous model in Simulink and the discrete model in SystemC which are schematic and textual respectively. The flow output is the global simulation model.

7.1 Simulink Interfaces

Simulink interfaces are classified into four types. They do not change if the synchronization mode changes and can be parameterized from their dialog box. The interface types are:

Inter_in implements the communication layer (input function), the "context switch" layer and a part of the "discrete events detection" layer which is responsible for detecting the passage of the solver by the discrete event time stamps (signal update events) and synchronizing them. Its parameters consist of the number of input signals.

Inter_out implements the communication layer (output function), the "context switch" layer and a part of the "discrete events detection" layer which is responsible for detecting the passage of the solver by the discrete event time stamps (sampling events) and synchronizing them. Its parameters are identical to those of the Inter_in.

Inter_state implements "detection and sending of state events" as well as the "context switch" layer. Its parameters are the state event numbers.

Sync implements the remainder portion of the "discrete events detection" layer. It is responsible for creating break points which the solver (a variable step solver) must reach with accuracy. These break points are the time stamps of the received events (signal update events and sampling events). It does not have any parameters.

The Simulink interfaces are functional blocks programmed in C++ using S-Functions. These blocks are manipulated like any other components of the Simulink library. They contain input/output ports compatible with all model ports that can be connected directly using Simulink signals. Users start by dragging the interfaces from the interface components library into their models windows. These items are then parameterized and finally connected to the inputs and outputs of users' models. Before the simulation, the functionalities of these blocks are loaded by Simulink from the .dll libraries (Figure 6.15).

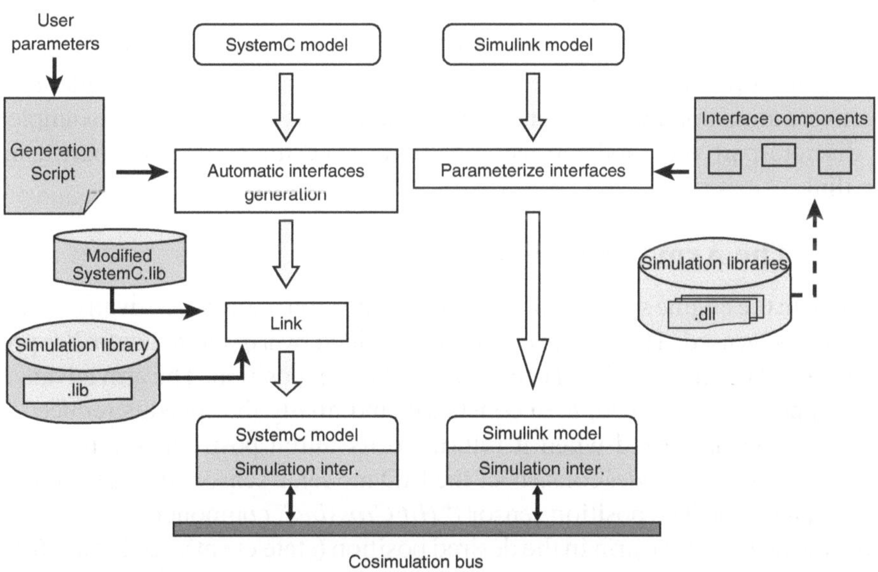

Figure 6.15. Flow of automatic generation of the simulation interfaces

7.2 SystemC Interfaces

For SystemC, as indicated above, some of the synchronization functionalities have been implemented at the scheduler level (which is a part of the state event management and the detection of the end of the discrete simulation cycle).

The interfaces are classified in the following fashion:

InterIn implements the communication layer (input function), a part of the "state event consideration" layer and the "context switch" layer (UE mode). It ensures synchronization with input data thanks to the sampling clock events (intern events). It can be viewed as a sampler circuit and contains its own parameters: (1) the names, numbers and data type of input ports, (2) the sampling periods, and (3) the mode used (e.g. FS, PE).

InterOut implements the output communication function and additional synchronization functionalities in the case of the UE mode. Its parameters are: (1) the names, number, and data type of output ports; and (2) the mode used.

Figure 6.16 presents an example of the "InterOut" interface with the UE Mode:

The interfaces are automatically generated by a script generator which is equipped with parameters defined by the user's input. Once the interfaces are generated, their connection is made within the function *sc_main*. The model is compiled and the link editor calls the SystemC library and a static library, called a "simulation library" (see Figure 6.15).

8. Experimentations

To analyze the capabilities of the continuous/discrete simulation model and its implementation, two illustrative examples were used: an engine controller used to activate a manipulator arm, and a sigma/delta converter. For both examples, the discrete part was modeled using SystemC and the continuous part using Simulink.

8.1 The Arm Controller

To regulate the engine speed, a closed-loop proportion, integral, derivative (PID) controller was used. The engine control is provided by a *discrete controller* providing speed orders calculated according to the arm position. The arm advances first at a progressive speed, then constantly and finally the speed is reduced. It runs at a constant speed when it returns to its initial position (Figure 6.18). The *continuous submodel* consists of the PID model, a sensor, an engine model and an integrator. The position sensor ("*Hit Crossing*" component) was used to signal the arrival of the arm in the desired position (state event), see Figure 6.17.

In order to evaluate the different synchronization models implemented in the presented simulation tool, the transaction level and ISA level execution

```
InterOutUE.h
#include "systemc.h"
#include "cosim.h"
SC_MODULE(interOut)
{
        sc_in<double> data;  //signals can be double or bit vector
        sc_in<double> data1;
        void send_data();

        SC_CTOR(interOut)
        {
                SC_METHOD(send_data);
                sensitive << data ;
                sensitive << data1;
                dont_initialize();
        }
};
InterOutUE.cpp
#include "InterOutUE.h"
void interOut :: send_data()
{
        // use the WriteSignalToSimulink function to send the time stamp of
        //signals update events
        WriteSignalToSimulink (sc_simulation_time()/1000000000, 100);

        // possible signal conversion
        .........
// send signals values
WriteSignalToSimulink (data.read(), 0);
        WriteSignalToSimulink (data1.read(), 1);
        // indicate the presence of new events
SwitchContextFlag ();
}
```

Figure 6.16. Excerpt of SystemC interface

models (presented in Chapter 3) were considered for the discrete part of the arm controller. At the first level, timing annotation functions were used to compute the time necessary since the system provides feedback. At this level, the system was simulated using the FS mode. For the ISA level, a SystemC implemented instruction set simulator for the DLX processor was used. The processor frequency was fixed at 4 MHz and the sampling period was fixed at 0.4 s. At this level, the UE mode with state event consideration (see Section 6.5.2.2) and the FS mode were used.

Table 6.3 presents the simulation time for an arm controller simulated during a period of 60 s.

Table 6.3. CPU time for the arm controller simulated for 60 s

Abstraction level for the discrete submodel	ISA level		Transaction level
Synchronization mode	FS	UE	FS
CPU time	25 min	5.31.s	0.26 s

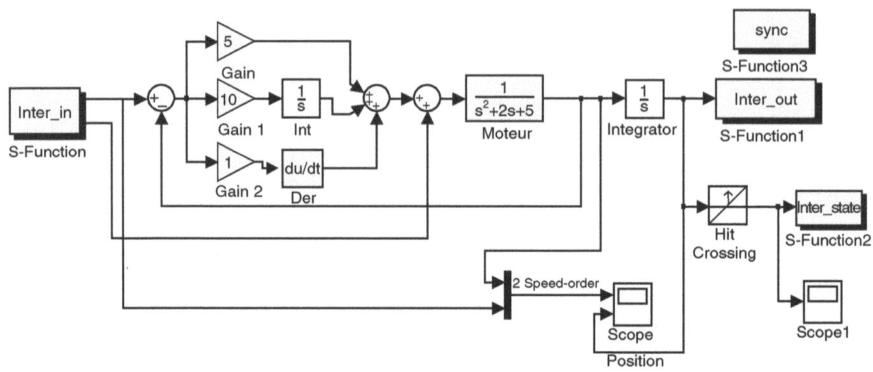

Figure 6.17. The continuous subsystem

Accuracy analysis. The accuracy analysis for the FS mode is presented. In the above example, the arrival of the arm at the desired position is a state event that occurs at the moment 15.4990 s. The *Inter_state* interface (Figure 6.17) signals the presence of this event to SystemC. Then the discrete controller sends a request to reduce the speed to zero. The state event was considered with accuracy by the discrete part and the passage of the order to zero was completed at the state event time stamp. These results are illustrated in Figures 6.18 and 6.19.

Figure 6.19 shows that SystemC scheduler planned an event $e1$at time 15.6 s while the arrival of the state event will force the scheduler to increment its time 15.4990 s. As shown in the figure, the $e1$ event, which is a clock event, will be treated after the state event. Figure 6.18 also shows the synchronization and data exchange accuracy. The continuous submodel sets the order to 1.5 at the time stamp with the value 3.2001 s, the moment it was modified by the discrete controller.

8.2 Σ/Δ Converter

Sigma-delta (Σ/Δ) converter is an over sampling analog-to-digital conversion technique. The analog input is over sampled N times faster than the requested digital output frequency and quantified by one bit, ± 1. The quantified value is fed back to the analog section, refined by an average filter and collected by a digital accumulator. For every Nsample, the converter produces the digital

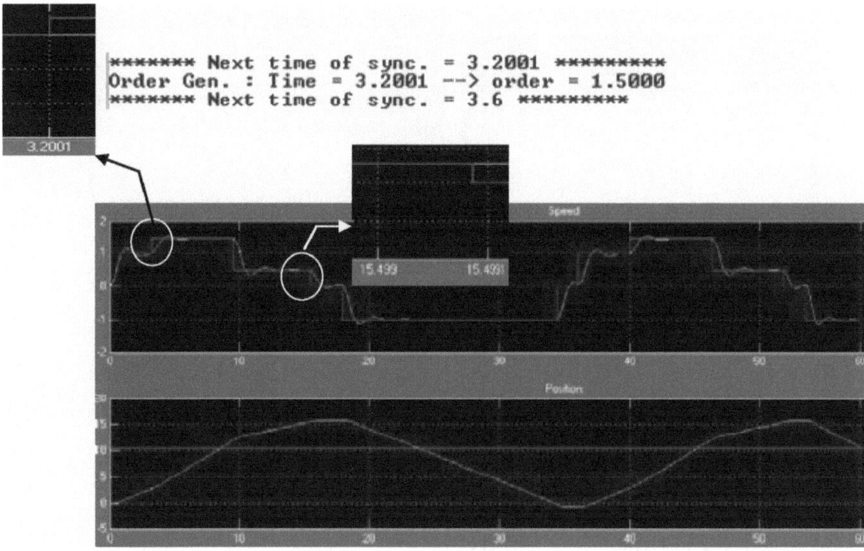

Figure 6.18. Speed, speed order and position (from Scope in Figure 6.17)

```
******* Next time of sync. = 15.6 *********
State event at time 15.4990
Order Gen. : Time = 15.4990 --> order = 0.0000
******* Next time of sync. = 15.6 *********
```

Figure 6.19. State event consideration (discrete controller output)

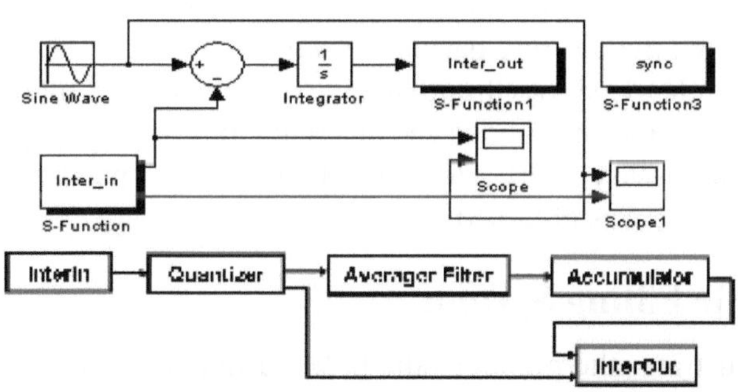

(a) Continuous model **(b) Discrete model**

Figure 6.20. Σ/Δ converter overview

output and resets the accumulator (see Figure 6.20). For this example the sampling frequency was 5.12 Khz. Figure 6.21a shows the modulated signal (scope) and Figure 6.21b shows the input and the digital output signals.

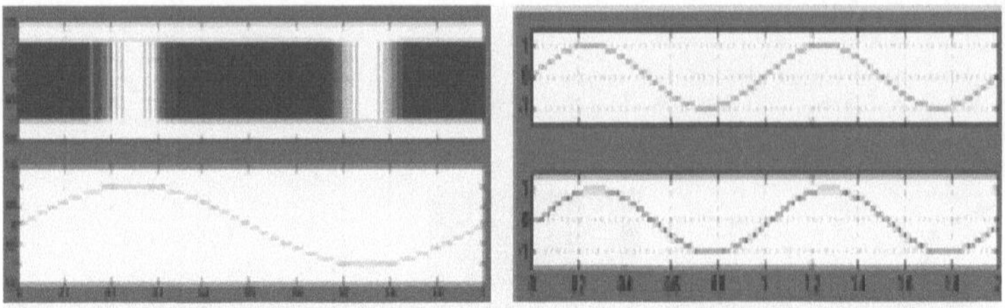

(a) Σ/Δ modulation (b) input and output

Figure 6.21. Σ/Δ converter signals

Table 6.4. CPU time for Σ/Δ converter

Application	Σ/Δ converter
Simulated time	2s
CPU time	0.8s

Table 6.4 presents time results for the converter simulation. This model simulation was conducted with UE mode since its continuous submodel does not generate state events.

Result discussion. Table 6.3 shows the acceleration of the FS model at the transaction level vs. the FS model at the ISA level for the discrete part of the arm controller. It is important to note that the discrete model at ISA level presents a better accuracy. However, the accuracy of the data exchange between the continuous and the discrete part is the same.

Table 6.3 also clearly shows the advantage of using the UE mode compared to the FS mode (where the continuous and discrete models are tightly synchronized resulting in unnecessary overhead) at the ISA level: a speed-up of about two orders of magnitude was obtained.

8.3 Bottle Filling System

Figure 6.22 illustrates the model of the bottle filling system of the system that should respect the following specification. The jobs (bottles) are externally generated where the generation process may be random. These jobs are queued in a "job buffer" (an FIFO). If there is no job in "setup" or in "job in process", the valve v1 is ON and the first job in the "job buffer" proceeds to "setup". The "setup" process model the delay required forn the positioning of the bottle at the right spot. After "setup" is completed, the job is placed in "job in process". When this happens, v2 and v3 are notified so that the actual physical process that defines the job can start. The "time-driven process", in the continuous model,

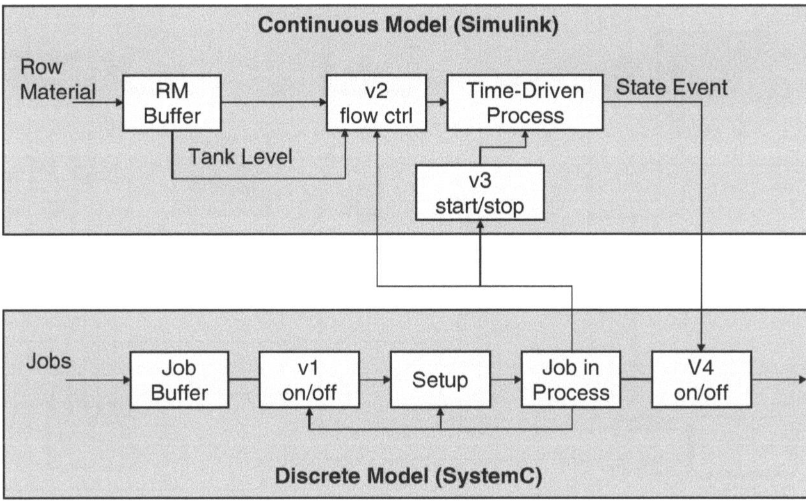

Figure 6.22. The bottle filling model

consists in filling an initially empty bottle with the raw material (RM) fluid to a given level (10 lit). It is activated by valve v3 as soon as a job is ready to start. The valve v2 acts as a controller of the bottle filling flow in the "time-driven process", which can have three values: 0 if there is no job in the "job in process", the "RM buffer" inflow (0.025 l/s) if the last one is empty and 0.033 l/s if not. When "time-driven process" is completed (the bottle level reaches 10 l), it sends a state event to valve v4 to open and let the current job to leave. At this time, "the job in process" opens v1 to accept the next job and after a small delay (the time to react to the state event), it signals v2 to stop the RM flow and v3 to reset the "time-driven process". For this application, the "setup" time is set to 60 s and the initial "RM buffer" level is set to 3 l. We give in Figure 6.23 an overview of the continuous model and its simulation interfaces.

Simulation results. The simulation was performed using the FS synchronization mode, since the continuous model generates state events and the signals update events are not periodic.

For the results given by Figure 6.24, the job arrival rate is set to 180 s. In this figure the outflow represents the bottle filling flow. The filling process starts with a 0.033 l/s. Each time the tank level is equal to zero, the outflow is switched to 0.025 l/s. The hit signal represents the state events generated when a bottle level reaches 10 l. The discrete model reacts to these events by switching the control signal to zero. Once a job is presented in "job in process", the DM switches its value to "1 in order to open v2 and v3. The signal 'job buffer', from the discrete model, represents the number of accumulated job in the "job buffer". This signal is sent to Simulink just for viewing purpose. We remark, in Figure 6.24 that

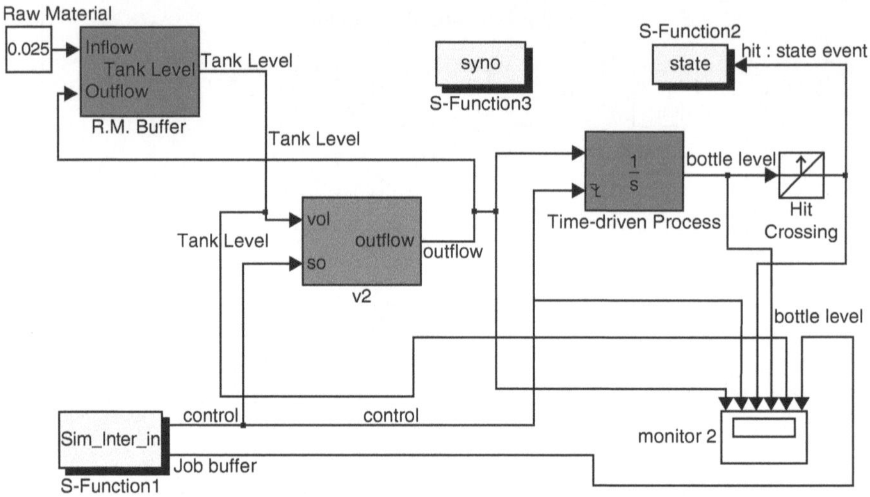

Figure 6.23. The continuous model overview

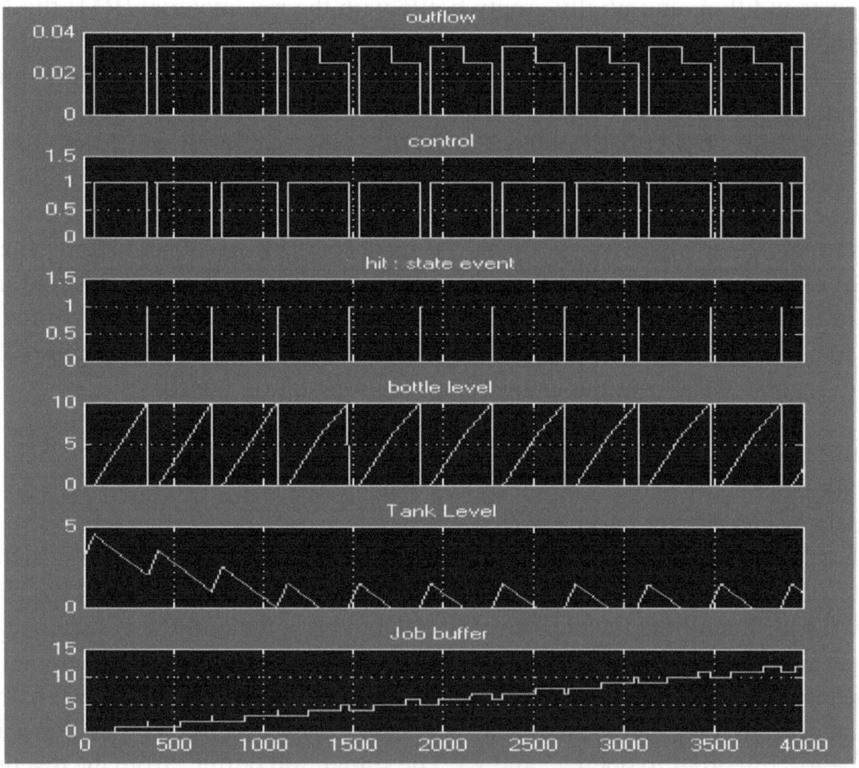

Figure 6.24. Simulation results

the number of accumulated jobs is increasing. Our experimentations shows that if jobs are accumulated in the "job buffer" (more than one job) then we risk usually to exceed the buffer capacity. To avoid this situation, the job arrival rate must be superior or equal to 400 s. This value depends on the outflow, the inflow and the "setup" time.

9. Conclusion

This chapter presented the anatomy of a discrete/continuous global simulation model. The first section presented the concepts manipulated by both models, the time distribution model as well as event management. In the second part, based on previous studies, several synchronization models that resulted from a deep analysis of synchronization issues with respect to accuracy and performance constraints were introduced. In the third part, the architecture of a generic global simulation model was proposed, providing semantics for the accurate global validation of discrete/continuous systems. It allows for the use of powerful tools in both domains. The global simulation model was implemented by simulation interfaces in order to produce global simulation model instances for discrete-continuous systems simulation using SystemC and Simulink. Finally, to evaluate the proposed simulation model, co-simulation results from two discrete/continuous applications were illustrated.

References

[Bal03] Balarin, F. et al., "Metropolis: An Integrated Electronic System Design Environment", *Computer*, vol. 36, issue 4, April 2003, pp. 45–52.

[Cel06] Celoxica, http://www.celoxica.com/methodology/

[Cal91] Callier F. M., Desoer C. A., *Linear System Theory*, Germany, Springer-Verlag, 1991.

[Cha96] W.T. Chang et al., "Heterogeneous Simulation – Mixing Discrete-Event Models with Dataflow", RASSP special issue of J. on VLSI Signal Processing, 1996.

[Fle95] J. Fleischmann et al., "Comparative Analysis of Periodic State Saving Techniques in Time Warp Simulators", Parallel and Distributed Simulation, 1995.

[Fre00] P. Frey et al., "Verilog-AMS: Mixed-Signal Simulation and Cross Domain Connect Modules", Behavioral Modeling and Simulation Workshop, 2000.

[Gea84] C. W. Gear et al., "Solving Ordinary Differential Equations with Discontinuities", *ACM Transaction on Mathematical Software*, vol. 10, 1984, pp. 23–44.

[Gup85] K. Gupta, et al., "A Review of Recent Developments in Solving ODES", Proc. of CSUR, vol. 17, No. 1, 1985.

[IEE99] *IEEE Standard VHDL Analog and Mixed-Signal Extensions*, IEEE Std 1076.1-1999, 23 December 1999.

[ITR06] International Technology Roadmap for Semiconductor Design, 2003.

[Jan04] Jantsch A., *Modeling Embedded Systems and SOCs*, USA, Morgan Kaufmann, 2004.

[Liu02] J. Liu et al., "On The Causality of Mixed-Signal and Hybrid Models", Hybrid Systems: Computation and Control, 2003, pp. 328–342.

[Mat06] Matlab-Simulink, www.mathworks.com

[Mar02] Martin D. E et al., "Integrating Multiple Parallel Simulation Engines for Mixed-Technology Parallel Simulation", Simulation Symposium, 2002.

[Mod06] Modelica, www.modelica.org

[Mod06] ModelSim® SE Foreign Language Interface, version 5.7d.

[Nic02] G. Nicolescu et al., "Validation in a component-based design flow for Multicore SoCs", in Proc. ISSS, 2002.

[Pat04] H. D. Patel et al., "Towards A Heterogeneous Simulation Kernel for System Level Models: A SystemC Kernel for Synchronous Data Flow Models", Proc. ISVLSI'04.

[Pto06] Ptolemy, http://ptolemy.eecs.berkeley.edu/ptolemyII

[Ree04] D. K. Reed et al., "An Application of Parallel Discrete Event Simulation Algorithms to Mixed Domain System Simulation", DATE'04, pp. 1356–1357.

[SDL06] SDL RT, specification & description language – real time, available at http://www.sdl-rt.org

[Sys03] SystemC LRM, 2003, available at www.SystemC.org

[Tah93] El Tahaway et al., "VHDeLDO: A new mixed mode simulation", DAC Conference, 1993, with EURO-VHDL'93. Proceedings EURO-DAC'93.

[Vac03] A. Vachoux, et al., "Analog and Mixed Signal Modeling with SystemC", Circuits and Systems, ISCAS'03.

[Val95] C. A. Valderrama et al., "A unified model for co-simulation and co-synthesis of mixed hardware/software systems", Proc. 1995 European Conference on Design and Test.

Chapter 7

Methodology for Heterogeneous Systems Validation

G. Nicolescu[1] and A.A. Jerraya[2]
[1]*Ecole Polytechnique de Montreal, QC, Canada,* [2]*CEA-LETI, Grenoble, France*

1. Introduction

The two anatomies presented in the previous chapters show that global validation of heterogeneous embedded systems requires very sophisticated execution models. These models include the execution of various components, the interpretation of the interconnects between the components as well as the adaptation of various execution models, abstractions, and communication protocols that are specific to the different components of the systems to be validated. This makes the design of these global execution models extremely challenging; it requires an increasing fraction of the overall design effort and represents a possible source of errors in this process.

In this context, the automatic generation of global execution models is essential for the design of heterogeneous embedded systems. As explained in Chapter 3, several works have been done in this direction. Unfortunately these solutions are appropriate for a given application domain: they are well adapted for certain steps of the design flow or they manage only the adaptation of certain abstraction levels and/or communication protocols.

The key issue for the definition of a generic approach is the definition of the internal architecture of the global execution model. This architecture gives the types of elements that have to be customized and combined in order to obtain a global execution model of a given system. These elements will compose a library for automatic generation of global execution models. Based on these remarks, this chapter presents a methodology for the validation of heterogeneous embedded systems.

The chapter is organized in six sections. Section 2 gives a general view of the flow for automatic generation of execution models. The virtual architecture

G. Nicolescu and A.A. Jerraya, Global Specification and Validation of Embedded Systems – Integrating Heterogeneous Components, 109–145.

concept and the generic internal architecture of a global execution model are introduced in Section 3 Section 4 presents the automatic generation of global execution models. The methodology is validated in Sections 5 and 6 by applying it to two heterogeneous systems: the optical switch and the VDSL modem. Section 7 concludes the chapter.

2. Flow for Automatic Generation of Global Execution Models

This section gives the general view of a flow for automatic generation of execution models. This flow, illustrated in Figure 7.1, is based on a *simulation library* and a *generic architecture of global execution model*. It enables the automatic generation starting from the virtual architecture of a heterogeneous system. The simulation library and the virtual architecture are inputs for the module in charge with the generation of the execution model. The output of this flow is the global execution model, with all its components introduced in Chapter 4: the simulators, the simulation interfaces, and the co-simulation bus. The virtual architecture concept and the generic internal architecture of the global execution model are presented in the next section.

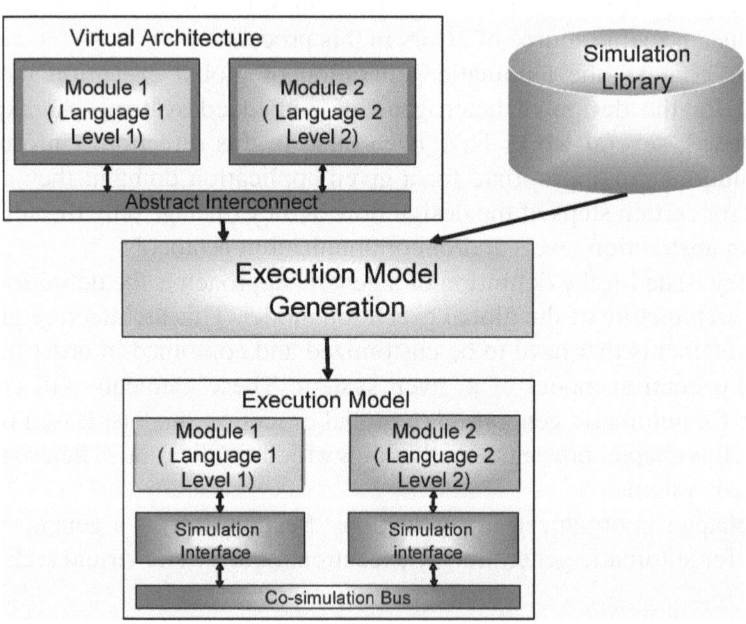

Figure 7.1. Overview of a flow for automatic generation of execution models

3. Virtual Architecture of an Embedded System and Global Execution Model

3.1 Virtual Architecture of a Heterogeneous System

The *virtual architecture* is a nonexecutable representation of heterogeneous systems and represents the input in the flow for the generation of execution models. The basic idea in defining the virtual architecture is the separation between the communication and behavior.

The first concept allowing the separation of the two notions in the specification of electronic systems is the *wrapper of a module*. This concept represents the abstraction of the interconnection between two heterogeneous components and allows the separation between the module's behavior and the communication. The wrapper is a set of two types of ports:

- *Internal ports*, which are the ports of the module
- *External ports*, which are the ports allowing the connection with the communication channels

The different internal or external ports of a wrapper are put in correspondence and grouped into *virtual ports*. A virtual port can include m internal ports and n external ports (where n, m are natural numbers). The internal and external ports of a *virtual port* can be specified at different abstraction levels or using different specification languages. They can also be assembled to create a *hierarchical port*.

A *virtual module* is composed of a module and its wrapper. The different channels of communication connected to a virtual port can be grouped in *virtual channels*.

The Figure 7.2 shows an example of how the presented concepts can be used. The Figure 7.2a shows a module that has to be interconnected with communication channels at different levels of abstractions (e.g. a module at the transaction level that has to be connected with physical signals). Figure 7.2b shows how the connection between these heterogeneous components was abstracted and specified using the concepts of *virtual port*, *virtual channel*, and *virtual module*.

Using these concepts, a heterogeneous system can be modeled by a *virtual architecture* as a set of interconnected virtual components. The internal and external ports of the modules' wrappers can be heterogeneous (in terms of specification language, abstraction levels, and communication protocols). The wrappers represent an abstraction of the behavior that is necessary for the refinement or the execution of the virtual architecture. Therefore, the virtual architecture has to allow the annotation with configuration parameters. Figure 7.3 shows an example of virtual architecture.

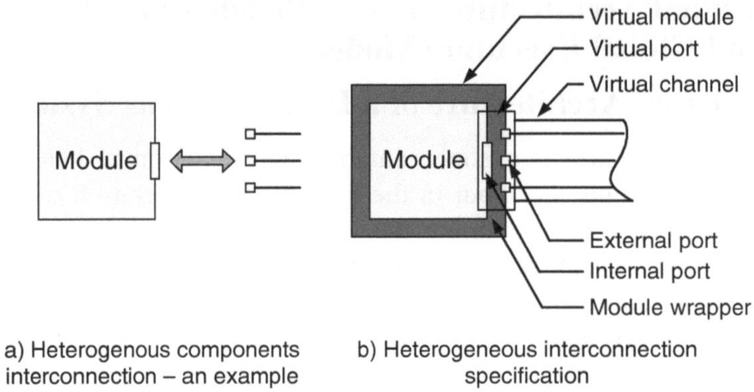

Figure 7.2. Basic concepts for the heterogeneous specification

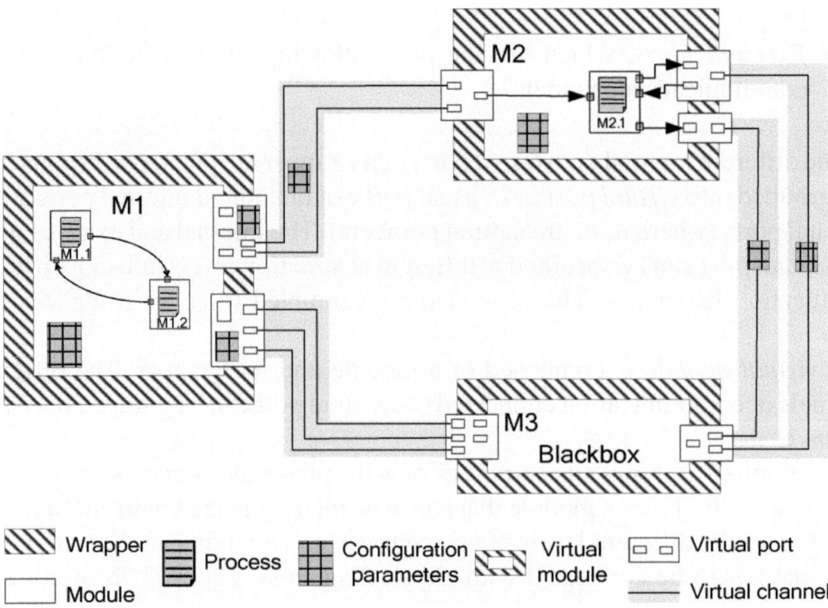

Figure 7.3. Virtual architecture model

In Figure 7.3 we observe three modules M1, M2, M3 and their wrappers. In the present example M3 is considered to have a different specification language (that was represented here as a blackbox). An important feature that is illustrated in this example is the parameters configuration. The ports, the modules, and the communication channels can have their parameters configured by user.

3.2 Generic Architecture of Global Execution Model for Heterogeneous Systems

This section presents a generic architecture for global execution models for heterogeneous systems. To define a global execution model, the designers have to first define the co-simulation interfaces structure. The co-simulation interfaces are in fact the execution models of the wrappers of different modules (see Figure 7.4)

The starting point in the definition of this model is the execution model used generally for the co-simulation: the different modules forming a system communicate through the co-simulation bus, via co-simulation interfaces. This model was introduced in Chapter 4.

The co-simulation interface has a very complex functionality that is to ensure the adaptation of a module to the co-simulation bus. It is mostly the adaptation of simulators, the synchronization and the adaptation of the communication.

To separate the synchronization from adaptation of the simulators and the communication, we propose a first partition of the co-simulation interface in three interfaces (see Figure 7.5b):

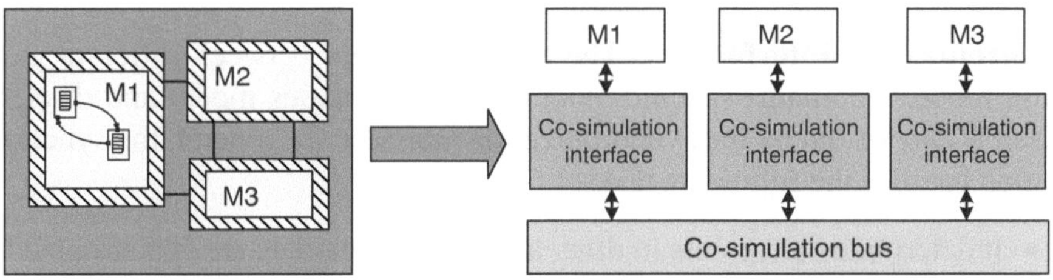

Figure 7.4. Heterogeneous specification and corresponding execution model

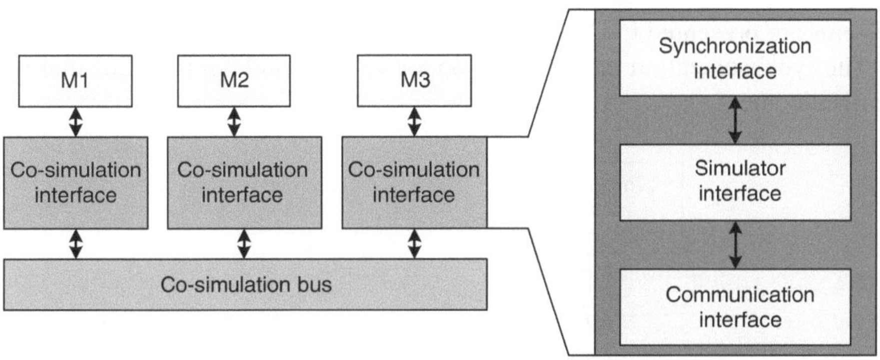

Figure 7.5. Separation of synchronization, simulator and communication adapters

- *Synchronization interface*, which is in charge with the synchronization between different execution models (such as Continuous/Discrete models)
- *Simulator interface*, which is in charge with the adaptation between the simulator and the co-simulation bus
- *Communication interface,* which is in charge with the adaptation of different communication protocols or abstraction levels

By separating the synchronization, the simulator adaptation and the communication the execution model becomes more extensible. The addition of a new simulator in the simulation environment needs only a new simulator interface and eventually, a new synchronization interface (if the user adds a new execution model). Also, the addition of new communication protocols or abstraction levels demands only the implementation of new communication interfaces. This separation also makes the execution model more flexible: the substitution of a simulator with a simulator of the same type (e.g. the replacement of a VHDL ModelSim simulator with a VSS simulator) does not require changing the whole co-simulation, only the adaptation of the simulator interface.

The next section presents the components of the co-simulation interface.

Synchronization interface. The *synchronization* between different execution models (domains specific models, e.g. continuous model and discrete model) is carried out by the synchronization interface. In general, the synchronization handles the following tasks:

- It determines the points in time, at which the models are synchronized.
- It determines the order, in which the models are simulated.

Figure 7.6 exemplifies the synchronization interfaces in a continuous/discrete execution model. More details on the synchronization of the continuous/discrete systems are presented in Chapter 6.

The synchronization interface sits on top of the simulator interface that will be detailed next.

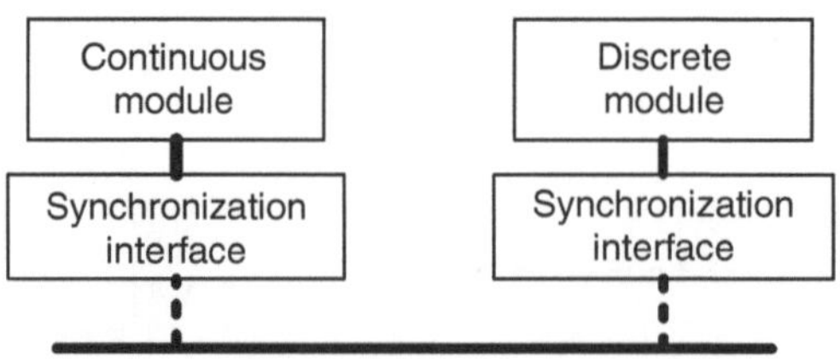

Figure 7.6. Example of synchronization interfaces for a continuous/discrete execution model

Simulator interface. In the global execution model, the *simulator interface* is in charge with the adaptation of a simulation environment. It makes visible the existence of an external simulator in the co-simulation instance. This type of interface has to ensure the accurate data exchange with the external environment. [Lem00] and [Cos99] present studies on the implementation of the simulator interfaces. This interface only solves the specification language problem of the heterogeneity of the internal and external ports forming the module wrapper.

Figure 7.7 shows a module and its simulator interface. They communicate through point-to-point interconnections. For the communication with the module that is integrated in the execution model, the simulator interface has ports that are dual for the module ports (e.g. an input port of the module implies an output port of the simulator interface). For the communication with the execution model, the simulator interface has the same type of ports as the module it adapts. The inputs/outputs of the simulator interface are given by the internal ports of the wrapper of the module to adapt.

The adaptation of the communication between the internal and the external ports of a wrapper is realized by the communication interface that will be presented in the next section.

Communication interface. The communication interface is necessary when an adaptation, in terms of abstraction levels or communication protocols is required. Figure 7.8 shows an example of a module that requires a communication interface. The abstraction level of this module is N (e.g. RTL) and it is interconnected with communication channels situated at the abstraction level M, with $N \neq M$ (e.g. the transaction level previously presented). For the representation we used the *virtual architecture* concept (see Section 3.1). Several

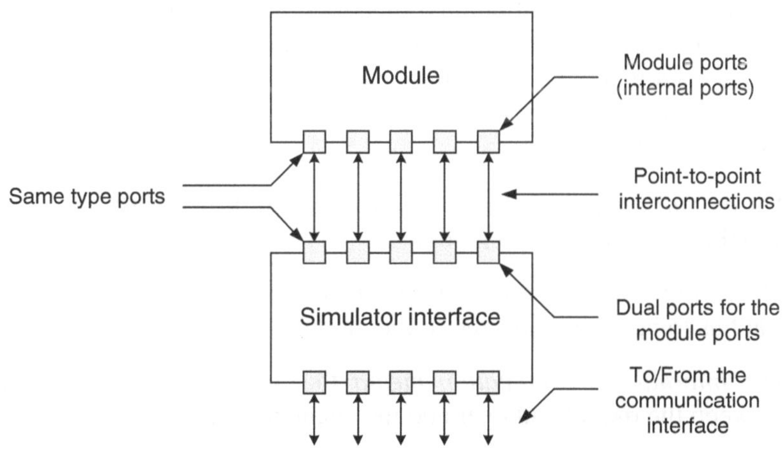

Figure 7.7. Module and simulator interface

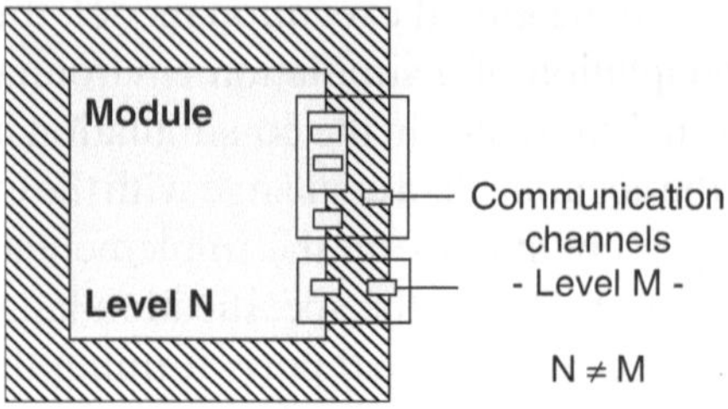

Figure 7.8. Example of a virtual module that requires a communication interface for the simulation

adaptations might be necessary between the internal and the external port of the wrapper:

- The internal and the external ports of the wrapper can be characterized by different communication primitives; also the module's behavior can call via the internal ports other communication primitives than the primitives provided by the communication channels.
- The internal and the external ports of the wrapper can manipulate different types of data.
- It is also possible that the adaptations require arbitration between the internal and the external ports. Thus, data can be sent simultaneously from several internal ports to an external one. Moreover, data can be sent simultaneously from several external ports to a unique internal port.

The communication interface is in charge with all these adaptations. To manage the complexity of this interface we propose its division in several different basic modules with very well-defined functions. The structure of the interface is presented in Figure 7.9. For a clearer representation, the figure does not show the synchronization and the simulation interfaces.

The communication interface has a modular internal structure formed by three basic elements:

- The *module adapter* that is specific to the module that has to be integrated into the global execution model
- The *channel adapter* that is necessary for each communication channel that has to be interconnected to the module
- The *internal communication media* that ensures the communication between the module adapter and the channel adapters

In terms of inputs/outputs, the communication interface has input and output ports that correspond to the wrapper of the module to be integrated in

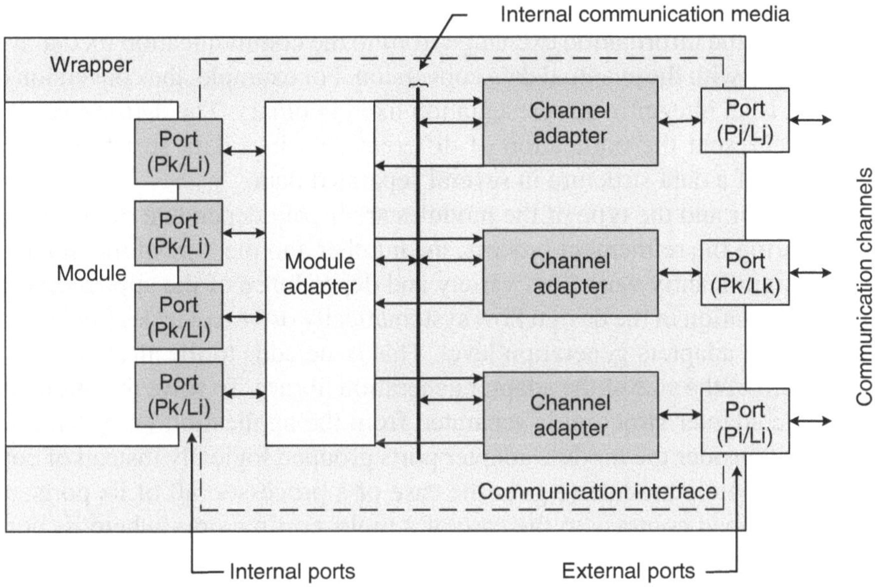

Figure 7.9. Generic architecture of the communication interface

the co-simulation instance. These different ports can be specific to different abstraction levels or different communication protocols. In Figure 7.9 the communication protocols (P) and the abstraction levels (L) are represented by *Pk/Li, Pi/Li, Pj/Lj, Pk/Lk*, and *Pl/Ll*. This structure allows the designers to separate the behavior and the communication and this is an important advantage. Also, the refinement (or the replacement) of a module implies only the choice of different module adapters. The *channel adapter* is not changed. The refinement at communication level of the channel will change only the channel adapter in question.

This structure allows not only the management of the complexity of the adaptation process but also reduces the number of possible adapters. Assuming the case when a module can be chosen between m possible types (considering also the different levels of abstraction) and where a channel is chosen from n possible channels, the numbers of adapters needed is $m \times n$ (if we assume that an adapter is a single block). Using the architecture presented above, the number of adapters that are required is $m + n$.

The following sections will describe the components of the communication interface.

Module Adapter. The *module adapter* is in charge with the module integration in a co-simulation instance. Its main functions are to supply the communication primitives called in the module's behavior via the internal ports and to ensure the information exchange from/to the communication media. It is also in charge with the eventual data conversion. For example, the conversion of a data that is an integer in a representation fixed in binary. The data conversion can also represent the unification of different data in a data structure or the partitioning of a data structure in several separated data.

The number and the type of the modules are highly dependent on the application. During the refinement process, the number and the type of the modules used can significantly vary. This variety and dependence of the application, as well as the position in the design flow systematically drive to a lack of flexibility at the module adapters generation level. This issue adds to the already important problem of the size of the adapter generation library. To solve this problem, the module adapter structure is separated from the application or system. We propose to consider the module adapter ports grouped logically instead of considering the whole module (e.g. in the case of a processor, all of its ports are logically grouped contrary to the case of a multi-port memory where its ports are independent from a logical point of view). Using this methodology, the diversity of modules is easily managed and the size of library for the adapters' generation is considerably reduced. In terms of input/output, the ports of the adapters can be grouped into two main categories (see Figure 7.9):

- The internal ports of the wrapper of the module to be adapted

- The ports that are specific for the internal communication media

As explained in Section 3.1, the ports can be grouped in hierarchical ports. The structure of the module adapter is in direct relation with the internal ports of the module wrapper, ports that are assembled in hierarchical ports. In this way, the module adapter will be split in elementary blocks where each block corresponds to a hierarchical internal port.

The module adapters can be constructed using elementary blocks that are named *port adapters*. The port adapter provides adaptation for the module ports that are adapted.

Port Adapters. A *port adapter* is in charge with a part of the functionality of the module adapter, mainly the part concerning the port that is adapted. This adapter supplies the communication primitives requested by the behavior of the module via the group of internal logical ports and ensures the data transfer "from" and "to" the internal communication media. It is also in charge with an eventual data conversion.

The principle for partitioning module adapters in port adapters is shown in Figure 7.10. For simplification, the channel adapters are not represented. On

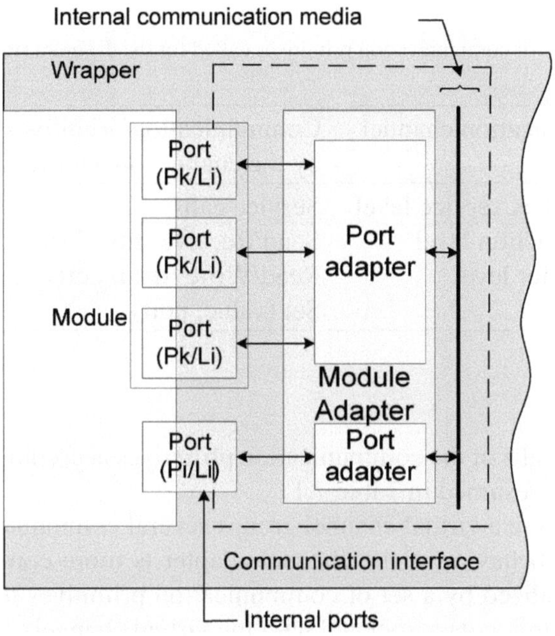

Figure 7.10. Module adapter as a structure of port adapters

the left side we represented a module with several ports. We can see that part of these ports is grouped to create a logic set or a hierarchical port.

Channel Adapters. The *channel adapter* is part of the communication interface and it ensures the interconnections adaptation in a co-simulation instance of a virtual architecture. In a virtual architecture the interconnections are represented by virtual communication channels. A virtual communication channel can encapsulate one communication channel or several communication channels logically grouped.

In the case where a virtual channel encapsulates only one communication channel, the channel adapter behavior consists only in transferring the information from/to the internal communication media. This is realized by calling the communication primitive supplied by the communication channel to be adapted and using the communication primitives specific to the internal communication media. If we consider the abstraction levels for the communication defined in Chapter 3, a channel adapter at service level calls a service, the channel adapter at transaction level calls *send/receive()* primitives, a channel adapter at transfer level calls primitives specific to a communication protocol (for example *fifo_write()*), a channel adapter at physical level execute set/reset on physical signals.

Table 7.1. Example of communication primitives called for the different types of communication channels

Type of communication channel	Communication primitives called by the channel adapter (example)
Abstract network at service level	Service call
Channel at transaction level	Send/Receive (file, disk)
Channel at transfer level	Read/Write (data, port)
Physical wire	Set (value, port)

The different calls of the communication primitives according to the channel to be adapted are resumed in Table 7.1.

In the case where a virtual channel groups several communication channels (e.g. a bus), the behavior of the channel adapter is more complex: the communication is realized by a set of communication primitives that respects the semantic of the protocols associated with the virtual channel.

Internal Communication Media. *Internal communication media* allows the exchange of data between the two types of adapters: module adapter (composed eventually of several port adapters) and channel adapters. The *internal communication media* guarantees an accurate interconnection between the basic components of the communication interface. It is also in charge with an eventual arbitration of the communication.

In Section 3.1, the virtual ports were presented as ports grouping internal and external ports that transfer the same information. The communication media has to take into consideration this aspect and has to ensure, jointly with the port adapters and channel adapters, the routing of the information between the internal and the external ports composing a virtual port.

More details on the different components of the communication interface required for the co-simulation instances generation are presented in Section 3.3.

Co-simulation bus. For the virtual architecture execution model, the co-simulation bus could have one of two possible functions depending on the following situations:

1. The designer of the virtual architecture does not provide an implementation of the communication. In this case, the co-simulation bus includes the execution model for the communication channels of the virtual architecture.

2. The behavior of the communication channels is provided by the designer in the virtual architecture. In this case, the communication channels are used as modules in a co-simulation instance of the virtual architecture. Thus, the co-simulation bus includes a set of signals ensuring a simple transfer of data or communication primitives calls between different elements of the co-simulation instance.

3.3 Virtual Architecture vs. Global Execution Model

Based on the definition of the virtual architecture and the execution model presented in Sections 3.1 and 3.2, this section shows the relationship between the different concepts of virtual architecture and execution model.

The *virtual architecture* was defined as a set of *virtual modules* interconnected by *virtual communication channels*. A *virtual module* is composed by a module from the heterogeneous system modeled by the *virtual architecture* and a *wrapper*. The wrapper can be composed of internal ports specific to the wrapped module and external ports specific to the communication channels connected to this module. The internal or external ports that are used for the transmission of the same information are grouped into virtual ports.

The *execution model* of a virtual architecture is a set of modules that are executed using different *simulators* interconnected through a *co-simulation bus via co-simulation interfaces*. Each co-simulation interface can be composed by a *synchronization interface*, a *simulator interface* and a *communication interface*. The inputs/outputs of the simulators interface are given by internal ports of the module wrapper. The communication interface is composed of a *module adapter* and several *channel adapters*. The module adapter is formed of several *port adapters*. The port adapter consists of internal ports put together in hierarchical ports. For each external port or group of external ports forming a hierarchical external port, we have a channel adapter. The channel adapters and the module adapters communicate through the *internal communication media*. The structure of the communication media is given by the virtual ports of a module's wrapper. As for the co-simulation bus, it can include the execution model of different communication channels of the virtual architecture (when the behavior is not provided by designers), or it is in charge only with the simple data transfer or communication primitive calls (when the behavior of these channels is already provided by the designers).

The correspondence between a virtual architecture and the structure of its execution model is given in Table 7.2.

For a better illustration, Figure 7.11 gives an example of a virtual architecture and its execution model.

Figure 7.11a introduces the virtual architecture for a system composed by three modules. These modules are situated at abstraction levels that are different

Table 7.2. Basic concepts of virtual architecture versus corresponding execution model

<table>
<tr><th colspan="2">Basic concepts for a virtual architecture</th><th>Basic concepts for the corresponding execution model</th></tr>
<tr><td colspan="2">Module</td><td>Simulator that executes the behavior of the module</td></tr>
<tr><td colspan="2">Wrapper</td><td>Co-simulation interface</td></tr>
<tr><td rowspan="4">Virtual port</td><td rowspan="3">Internal ports grouped (eventually) in hierarchical ports</td><td>Synchronization interface</td></tr>
<tr><td>Simulator interface</td></tr>
<tr><td>Port adapters composing a module adapter</td></tr>
<tr><td>External ports grouped (eventually) into hierarchical ports</td><td>Channel adapters</td></tr>
<tr><td colspan="2">Virtual communication channels</td><td>Co-simulation bus or simulator executing its behavior</td></tr>
</table>

from the abstraction level of the communication network. Thus, the internal and external ports of the modules' wrappers are ports at different abstraction levels. We assume that the specification language for module M3 is different from the language used for the specification of the rest of the system. In this case, the wrapper of the module M3 consists of internal and external ports that are specific to different specification languages and different levels of abstraction.

Figure 7.11b shows the execution model corresponding to the virtual architecture in Figure 7.11a. This is the first hierarchical level of the execution model. Each module is connected to the communication network through a co-simulation interface. The co-simulation interface for each component relates in fact to the execution model of the wrapper.

Figure 7.11c shows the details of the co-simulation interfaces that are necessary for the execution of the virtual architecture. For each wrapped module, for each internal hierarchical port of the wrapper there is a port adapter and for each virtual channel corresponds a channel adapter. M3 is specified in a different specification language (such as a continuous specific language) than the rest of the system described using a discrete simulator. Thus, a simulation interface is mandatory for the execution. The module also requires synchronization interface with the anatomy presented in Chapter 6. The co-simulation bus provides the behavior of the communication channels.

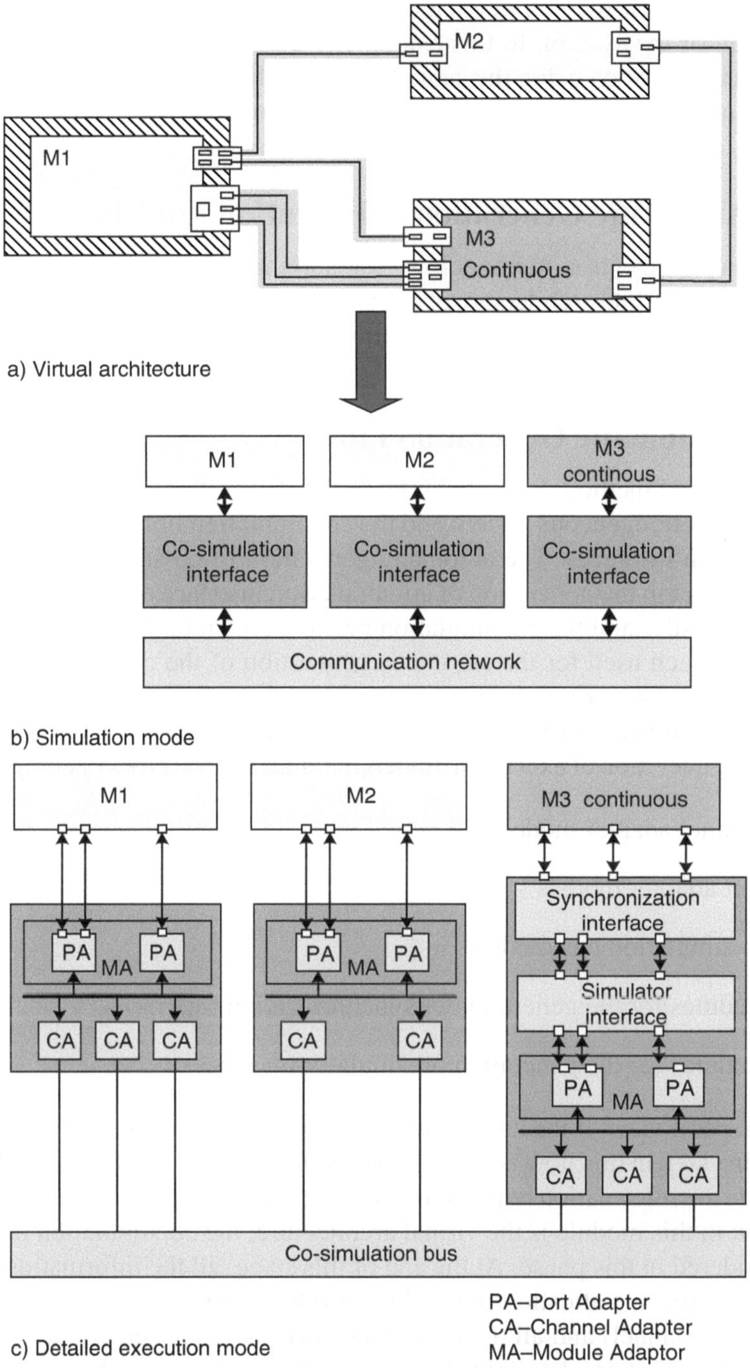

Figure 7.11. Virtual architecture and corresponding global execution model

The behavior of the communication channels was not provided in the first description of the system. In this example, we assumed that the modules M1 and M2 are described using the same languages that the language used for the simulation interfaces and the co-simulation bus in the global execution model of the system.

4. Automatic Generation of Execution Models

This section presents a methodology for automatic generation of execution models for the virtual architecture of a heterogeneous system. This methodology is based on the generation of co-simulation interfaces by assembling elements located in a co-simulation library.

4.1 Automatic Generation Flow

In the proposed methodology, the input in the design flow is the virtual architecture of a heterogeneous system and the co-simulation library is used to build co-simulation instances. The automatic generation of co-simulation instances consists in fact in the generation of co-simulation interfaces and their assembly with the modules and the co-simulation bus in a co-simulation instance.

The approach used for the automatic generation of the co-simulation interfaces is based on the principle of components configuration and assembly. These components already exist in a co-simulation library. The library used for the automatic generation of execution models includes several types of components:

- channel adapter models
- port adapter models
- co-simulation bus models
- modules for the generation of synchronization interfaces
- modules for the generation of simulator interfaces

The general view of the automatic generation flow is shown in Figure 7.12.

The first step in this flow is the *analysis of the virtual architecture*. Its goal is to provide the information required by the next steps in a directly usable format. The input in this module is the virtual architecture; the co-simulation library is not considered in this phase. At the end of this stage, all the information on the topology of the specification (e.g. which modules are connected with a communication channel) and all the adaptations necessary for the execution of the input specification are determined. These adaptations include the specification language for modules, the abstraction levels adaptations, the synchronization, the type, communication protocols, and simulation environments. Only this step is the in direct contact with the input specification, the virtual architecture; it will be the only that will change if the virtual architecture is replaced by other type of representation.

The second step in the flow is the *selection of the components from the co-simulation library and the construction of the co-simulation interfaces.* Using the results from the specifications analysis, appropriate components are extracted from the library and are configured accordingly with the needs determined during the analysis phase, based on the system's specification. Once these elements from the co-simulation library are extracted and configured, they are assembled together to build the co-simulation interfaces. This step provides the information on the co-simulation interfaces topology that is required for the next stage of the automatic generation of execution models flow.

The last stage of the generation flow is the assembly of the components (the co-simulation interfaces and the co-simulation bus) required for the execution model. In fact, during this stage the co-simulation interfaces and the co-simulation bus are included in the structure of the system. To accomplish this, the information on the virtual architecture topology that was determined during the analysis phase as well as the results from the components and interfaces integration phase are used as input for this step.

4.2 Application of the Validation Methodology for Embedded Systems Design

This section presents the application of the execution models and the automatic generation of execution models for the design flow of embedded systems. The section starts with the conceptual view of the design flow and afterwards presents the design flow and the automatic generation of execution models during this design flow.

Design flow – Conceptual View. The conceptual view of the design flow of heterogeneous embedded systems is shown in Figure 7.13. The input of the design flow, the virtual architecture, can be manually described or it can be automatically generated using systems analysis tools (for instance MCSE [Cal97]).

As shown in the Figure 7.13, starting from the virtual architecture, the RTL level representation is obtained through the generation of hardware/software interfaces that are in fact the implementation of the wrappers of the virtual modules. The virtual communication channels are transposed on a specific communication network (communication network at RTL level).

In order to allow the automatic generation of the representation on RTL level, the virtual ports and channels that form the virtual architecture are annotated by attributes that model the refinement decisions. These annotations allow the

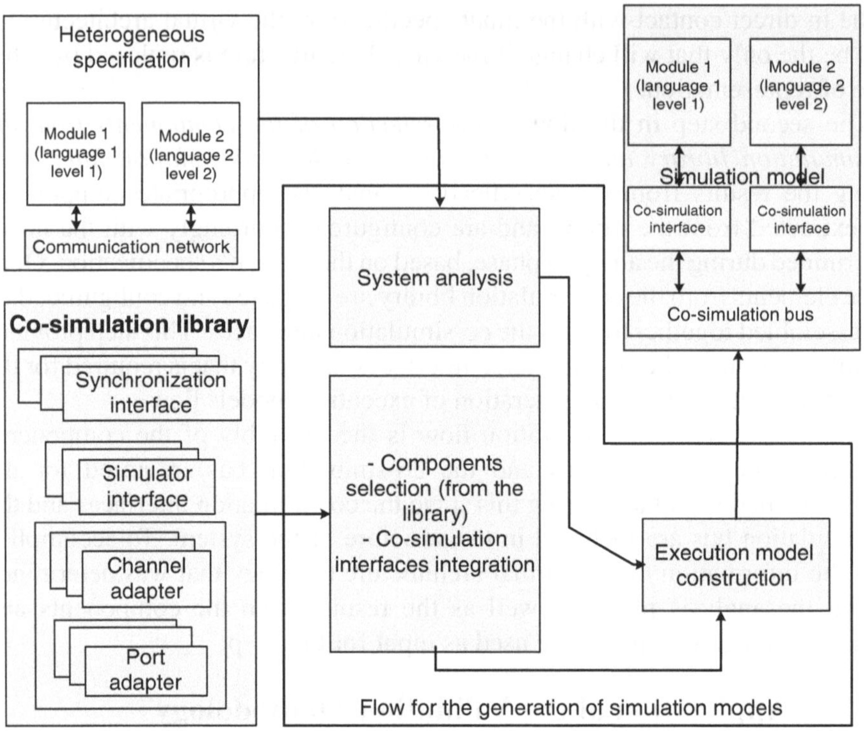

Figure 7.12. General view of the automatic generation of execution models

designer to specify the communication and computation resources needed for RTL level. This design flow ensures a refinement that is independent of components. The annotations for the execution are a must. These annotations allow the designer to specify the required adaptations in order to produce the execution model of a heterogeneous system.

Abstraction Levels for the Design Flow of Multiprocessor Systems. Figure 7.14 illustrates the abstraction levels for the presented design flow.

These abstraction levels were defined in Chapter 3. Ten abstraction level combinations are possible when these abstractions are considered during the design flow. A combination of abstraction levels considers two levels: one level from the communication and one level from the module hardware/software. Thus, there are six combinations for the software abstraction level, communication channels abstraction level and four combinations for the hardware abstraction level, communication channels abstraction level. All these combinations are presented in Table 7.3. Only three of these combinations do not require a communication interface for the execution: the combinations noted 1, 7, and 10.

All other abstraction level combinations (the combinations 2–6, 8, and 9 in Table 7.3) require a communication interface.

The automatic generation of global execution models allows the system's validation during different phases of the design flow by automatically generating co-simulation instances. This implies the generation of the co-simulation bus, simulators interfaces and communication interfaces for each possible com-

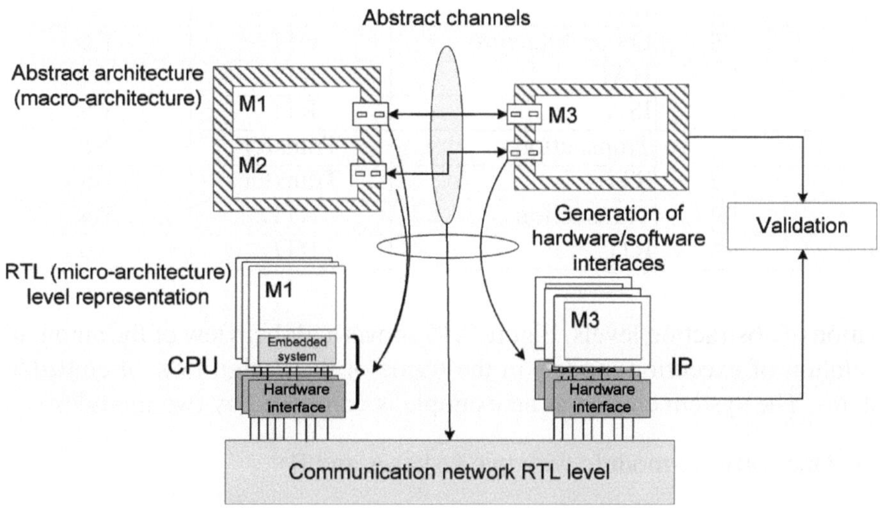

Figure 7.13. A conceptual view of the design flow of embedded heterogeneous systems

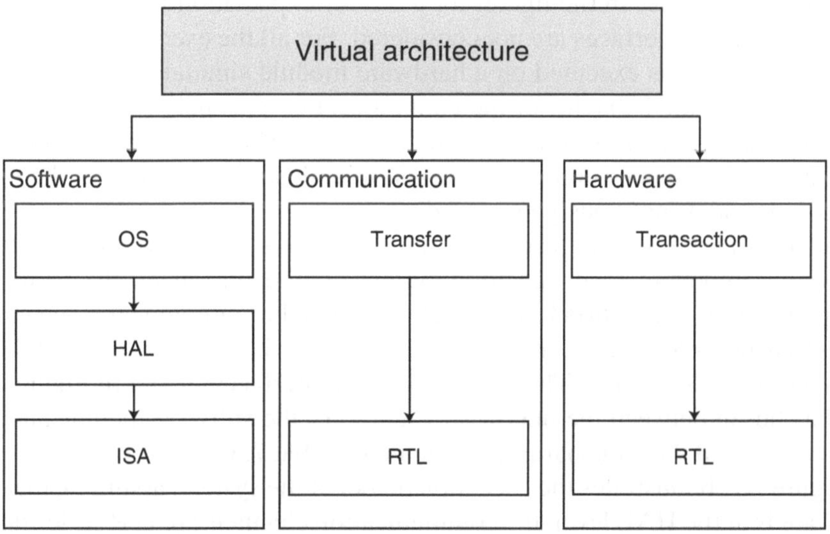

Figure 7.14. Abstraction levels in the considered design flow

Table 7.3. Abstraction level combinations in the design flow

Combination number	Module abstraction level		Communication channels abstraction level	Requires communication interface
1	software	OS architecture	Transfer	No
2		HAL	Transfer	Yes
3		ISA	Transfer	Yes
4		OS architecture	RTL	Yes
5		HAL	RTL	Yes
6		ISA	RTL	Yes
7	hardware	Transaction	Transfer	No
8		RTL	Transfer	Yes
9		Transaction	RTL	Yes
10		RTL	RTL	No

bination of abstraction levels. Figure 7.15 shows a global view of the automatic generation of execution models in the frame of the design flow of embedded systems. The system chosen as an example is composed by two modules:

- One software module with two tasks (A and B)
- One hardware module (C)

The main models of execution required for the validation are shown in the bottom of the figure. In the figure, for a clearer explanation the simulator and synchronization interfaces are not considered. For all the execution models the hardware module is executed on a hardware module simulator (e.g. SystemC or VHDL) and the behavior of the communication channels is executed on a simulator used to carry out the co-simulation bus. The type of simulator used for the software depends on each case and it is illustrated for each of the three execution models previously presented.

Figure 7.15a shows the execution model of the system, assuming that the software component is at operating system architecture level, the hardware component is at transaction level or RTL level and the communication channels are at transfer or RTL level. In this case the software is executed on a high-level simulator (e.g. SystemC). The communication interfaces shown in Figure 75a can be chosen between the interfaces 1 and 4 for the software module and the interfaces 8 and 9 for the hardware module in Table 7.3.

Figure 7.15b illustrates the execution model of the system, assuming that the software is at the HAL level, the communication channels are at RTL level and the hardware module is at transaction level or RTL level. In this case the software is executed natively on the host machine. In this case, an execution model for

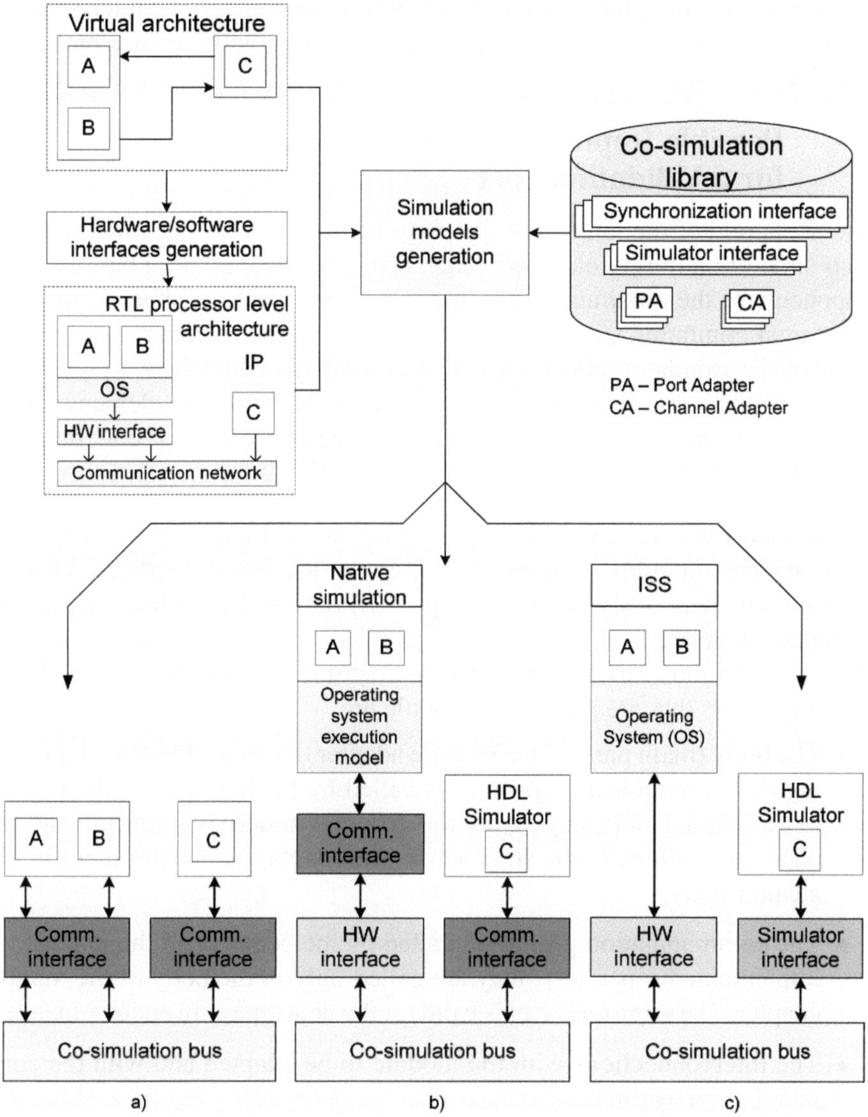

Figure 7.15. Automatic generation of execution models in the design flow of embedded systems

the operating system is necessary. Such model was introduced in Chapter 4. For the software module, the interfaces shown in Figure 7.15b correspond to the interface 5 in Table 7.3. For the hardware module the interfaces shown correspond to the interfaces 8 and 9 in Table 7.3.

In Figure 7.15c, the software module is at ISA level, the communication channels are at RTL level and the hardware module is at transaction or RTL level. In this case the software module is executed by the processor simulators (ISS).

The communication interfaces illustrated here correspond to the interface 6 for the software module and the interfaces 8 and 9 for the hardware module in Table 7.3.

4.3 Possible Implementation Solution for a Validation Tool

SystemC is one of the languages that may be used for the implementation of the co-simulation library elements. Therefore, the co-simulation bus, the basic components of the communication interface (port adapters, channel adapters, and internal communication media) as well as the other elements for the generation of the simulator's interfaces are specified in SystemC.

In the following sections we will give a few details on the elements of the library and on the realization of the flow for the generation of co-simulation instances.

Communication interfaces library. The channel and port adapters from the library are modules implemented in SystemC. These modules have the functionalities previously defined. Figure 7.16 details the module adapter and the channel adapter.

Figure 7.16a shows the module adapter structure in the co-simulation library. The components that are part of the module are:

- The body (main part) of the module adapter that is in charge with providing the communication primitives called by the behavior of the module to be adapted and carrying out the different underlying conversions, the data conversions. In the strict sense, the adaptation is carried out by the adapter body;
- The communication primitives called by the behavior of the module are implemented separately; they are called only in the body of the module adapter. The same behavior is valid for the data type conversion functions;
- The interconnections with the module to be adapted and with the communication media.

Figure 7.16a gives also the internal structure of body of the module adapter. This is a module that encapsulates a SystemC process. This module has two types of ports:

- The hierachical port that gathers all the ports used for the connection with the module
- The ports used for the interconnections with the communication media

The channel adapter does not realize arbitration and does not provide the communication primitives; it is a simple SystemC process. Its structure is given in Figure 7.16b. For the internal communication media the communication

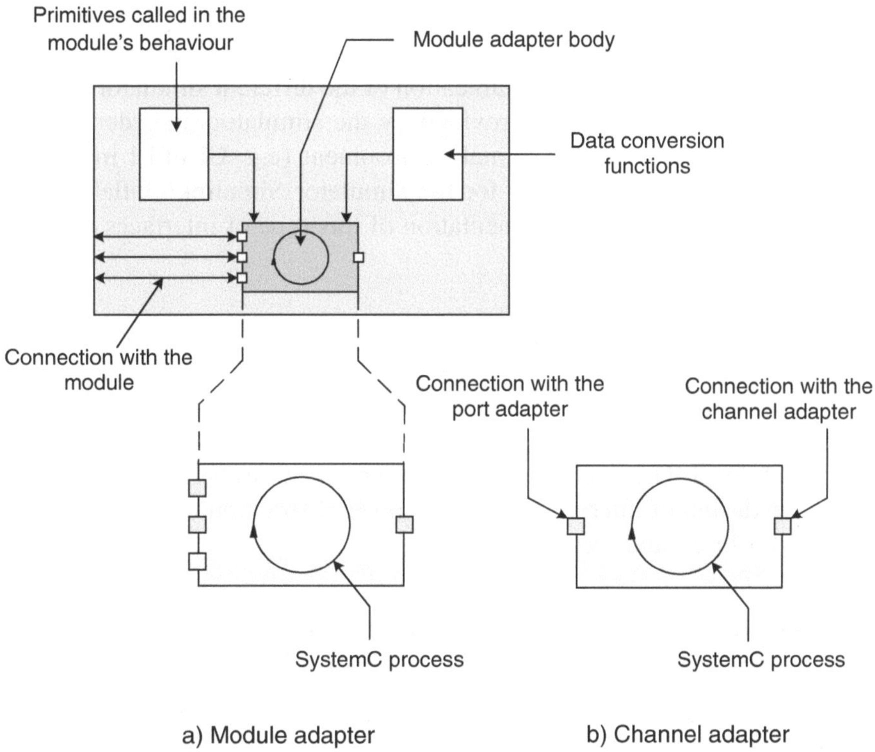

Figure 7.16. Communication library components structure

components provided by the SystemC library were used: communication by RPC, events, physical signal. The elements of the library are SystemC models that can be configured (e.g. the transferred data types, the size of a FIFO can be configuration parameters).

Synchronization interfaces library. In order to correctly interpret the information, the global execution model has to insure the synchronization between models. The synchronization library provides configurable components that allow the selection of synchronization mechanisms. These components are, in general, specific to the simulator. For instance, a possible solution for the continuous domain is to create a module that implements the synchronization function that creates break points which the solver (a variable step solver) must reach without going beyond [Mat02] while for the discrete domain, the synchronization can be implemented at scheduler level [Bou05]. More details about the continuous/discrete global execution model are presented in Chapter 6.

Simulator interfaces library. The elements for the simulator interfaces library are based on the SystemC library as well as on the UNIX library for the

Inter-Process Communication (IPC). Each simulator used during the simulation is in fact a process child of a SystemC process that communicates with its parent through IPC. For the communication of the different simulators through IPC we used different libraries provided by the simulators in order to allow the communication with the external environment (e.g. CLI/PLI in the case of the simulator VSS, S-function for the simulator Simulink/Matlab may be used). More details on the implementation of this type of interfaces are given in [Hes00].

Co-simulation bus implementation. The co-simulation bus is implemented in SystemC, like the libraries for the communication and simulator interfaces. The co-simulation bus has to be able to specify the communication at different abstraction levels. For instance, the signal and the ports provided by SystemC are used for RTL level. The interface and the communication channels concepts are used for the design of interconnections at lower abstraction levels (transaction level, transfer level, and service level).

5. Methodology Illustration for Validation of a Complex Heterogeneous System – an Optical Switch

Optical microsystems (MOEMS – Micro-Opto-Electro-Mechanical systems) integrate on the same chip different subsystems such as electronical, mechanical, and optical. In the past, these systems were only abstract ideas but now they are becoming ever more popular and they can be found on the market. The optical microsystems provide several advantages, compared with the classical switch: gain in size and speed, they are reliable and their design is becoming less costly [Wu97]. Thanks to these advantages they already promise to stir up optical networks with an optical switch that uses mirrors.

The heterogeneity of these systems imposes their modeling using different specification languages and simulators (e.g. C and HDL languages and simulators for the description of the electrical parts, Matlab/Simulink for the mechanical parts) and/or different abstraction levels. Thus, they are a very representative example to illustrate the methodology for specification and validation that was proposed in this chapter.

5.1 General Presentation

The general view of the optical switch is shown in Figure 7.17. The system is composed by three subsystems:

- The control subsystem that manages the movements of the mechanical mirrors that form the optical subsystem.

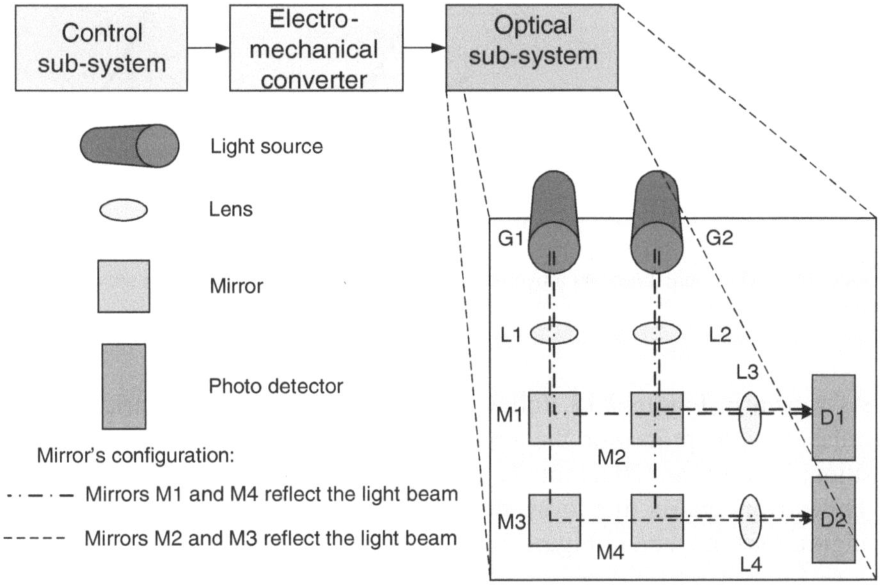

Figure 7.17. The optical switch

- The electro-mechanical converter that transforms the voltage sent by the control subsystem in mechanical commands (in terms of position for every mirror).

- The optical subsystem that is a 2×2 matrix, two light sources (G1 and G2 in the figure) and two photo detectors (D1 and D2 in the figure). Four lenses (L1÷L4 in the figure): L1 and L2 are collimators that limit the divergence of the light from the sources G1 and G2 and the lens L3 and L4 that focalize the light toward the photodetectors.

The sources and the detectors used during these tests are 3×3 matrices of basic beams sources/detectors. In this way, any combination of 3×3 light beams can be generated and/or detected.

The commutation is realized by the optical switch through mechanical movements of mirrors that are guiding the light flow from the entry sources (sources G1 and G2) towards the output (D1 and D2). The mechanical movement of the mirrors is controlled by the control subsystem. To ensure the commutation, the control subsystem controls the position of the mirrors. There are several ways of moving the mirrors and by consequence changing the optical configuration. The configuration is in this context a particular position of the mirrors. For this application, mirrors fixed at the end of an arm were used.

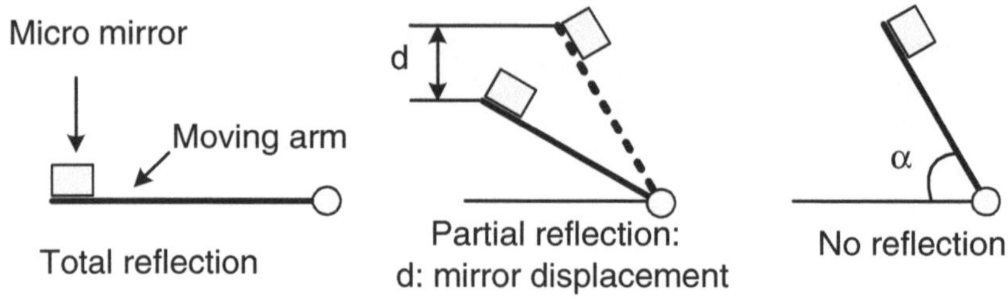

Figure 7.18. Mechanical mirrors movement comply with the commands from the control subsystem

As shown in Figure 7.18, following the intensity of the electrical current calculated by the processor, the arm can change its position. Depending on its position, mirrors can reflect totally, partially or do not reflect their inputs.

The mirrors matrix that forms the optical subsystem can be realized in two configurations (as seen in Figure 7.17):

- In the first configuration, the mirrors M1 and M4 totally reflect the beam while the mirrors M2 and M3 do not capture the beam. In this case the beam emitted by G1 is guided to the photodetector D1 and the beam emitted by G2 is guided to the photodetector D2.

- In the second configuration, the mirrors M2 and M3 totally reflect the beam while the mirrors M1 and M4 do not capture the beam. In this case the beam emitted by G1 is guided to the photodetector D2 and the beam emitted by G2 is guided to the photodetector D1.

The optical commutation is achieved by passing from one configuration to another.

5.2 Virtual Architecture

The virtual architecture concept was used for the global specification of the optical switch. Each subsystem interconnected through the virtual architecture was specified independently by a team. Thus, the design of the systems involved several teams. These teams cooperated only for the global functionality but none of them took into consideration, when designing their subsystem, the protocols/levels of abstraction or simulation environments used by the other teams.

The first team designed the control module using SystemC [Syn02], the second team implemented the electro-mechanical converter in Matlab/Simulink [Mat02] and the third team used Chatoyant for the optical subsystem. Chatoyant is a modeling and simulation environment for optical subsystems [Kur01].

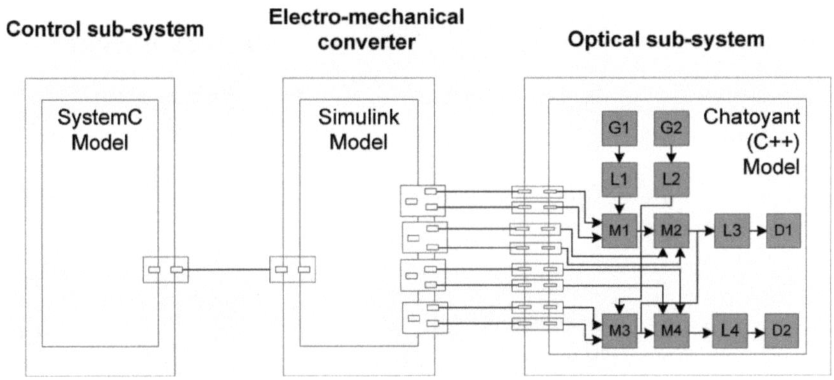

Figure 7.19. Virtual architecture of the optical switch

The electro-mechanical converter sends the mechanical commands by simple signals while in the mirrors modeling the mechanical commands are received through a simple "hand shake" protocol.

The virtual architecture of the optical switch is shown in Figure 7.19. Considering that the electro-mechanical converter is implemented in Matlab/Simulink, its wrapper includes internal ports specific to the simulation environment Simulink. These ports ensure a continuous communication through simple signals. For the connection with the rest of the system, the external ports are specific to SystemC where the communication is done at every clock cycle [Sys02]. As Figure 7.19 shows, each virtual port of the wrapper of the electro-mechanical converter connects an internal port of the converter with two external ports, these two ports being in fact the terminals of the "hand-shake" protocol used by the mirrors. Consequently, the model needs an interface for the Simulink simulator and a communication interface to adapt the communication protocols.

The virtual ports forming the wrappers for the control and the optical subsystems are constituted of ports of the same type.[1] There is no need to implement interfaces for these two modules in order to realize the execution model.

Execution model. For the validation of the virtual architecture of the optical switch, the corresponding execution model was implemented. As already explained in the previous section, the execution model contains a synchronization interface, a simulation interface, and a communication interface for the communication between the electromechanical converter and the rest of the system.

[1]To build the virtual architecture of the model it was necessary to encapsulate the sources in Chatoyant in SystemC modules, for each optical component. This step was done manually.

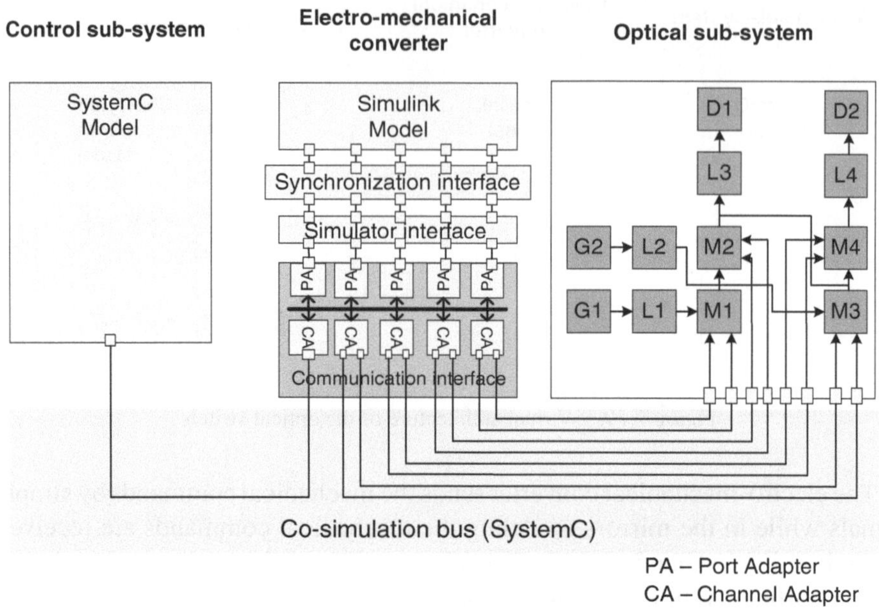

Figure 7.20. Execution model for the validation the optical switch

The execution model for the validation of the optical switch is shown in Figure 7.20.

5.3 Results

The reconfiguration of the optical switch was simulated. To observe the evolution of the system, the beams detected by the photodetectors D1 and D2 were displayed for each simulation step.

The sources/detectors used during the experiments could generate/detect all combinations of 3×3 light beams. To ease the results analysis, the two light sources were configured differently: the first source generates only a beam while the second source generates a combination of four light beams. We show the configuration in Figure 7.21 were we can see the nine squares representing the possible beams and the luminous points that represent the generated beams.

The results of the detections realized by the photodetectors D1 and D2 are presented in Figure 7.22 where every line of images corresponds to a detector.

In the initial state of the simulation, the mirrors matrix was in the first configuration. Consequently, after the first functioning step of the switch, the detector D1 detected the light produced by G1 and the detector D2 detected the light produced by G2. With the advancement of the simulation and after the commands sent by the control subsystem via the electromechanical converter, each

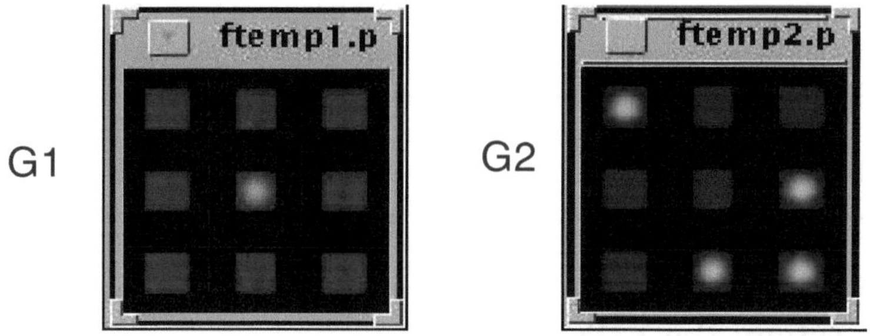

Figure 7.21. Configuration of light beam sources

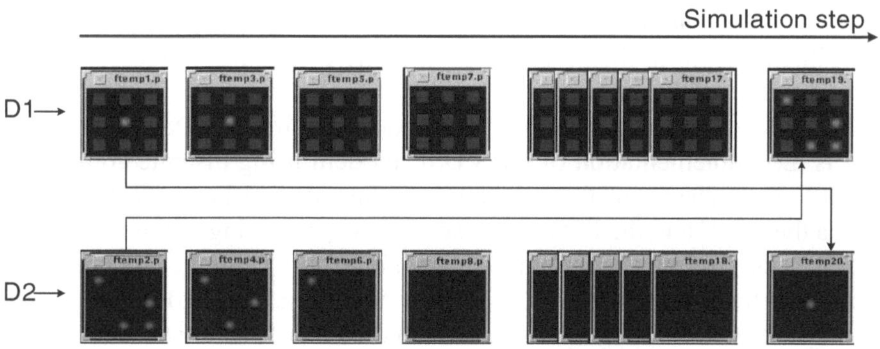

Figure 7.22. Optical switch reconfiguration

mirror gradually changed its position. The switch reconfiguration is also done gradually, in several functioning steps. After the second functioning step, mirror M4 slightly changed its position and now partially reflects the light produced by G2: it reflects three light beams from the four that were generated. Finally, after 11 functioning steps, the mirrors get into the position with total reflection (M2 and M3) or nonreflection (M1 and M4). As shown in Figure 7.22, at the end of the simulation, the detector D1 detected the light produced by G2 and the detector D2 detected the light produced by G1. The reconfiguration of the optical switch was therefore achieved. The co-simulation of the system was realized on a SUN Ultra-Sparc 1 machine and the simulation time was 30 s. That allowed a fast validation of the global functionality of the system, before the implementation of the final architecture.

These experiments required the communication between different teams, working on different subsystems composing the optical switch. One of the issues that emerged was that the designers of optical and mechanical systems were not used with high level concepts. On the other hand, one of the goals for the teams was to understand the whole semantic of the optical concepts used as well as the physical mechanics concepts. This problem was overcome by using the virtual architecture as a common reference for all the designers. Most of the time was used to enrich the co-simulation libraries and to encapsulate the C++ models of the optical components in the SystemC modules.

The global execution of the system allowed the improvement of the mechanical mirrors functionality as well as the global functionality of the system.

6. Specification and Validation for a Multiprocessor System: the VDSL Modem

6.1 General Presentation

The very-high-data-rate DSL (VDSL) is a communication protocol that uses the phone lines that are part of the digital subscriber line (xDSL) techniques. Given that it is still in the protocol stage, many companies propose their own versions of the VDSL protocol.

The application used in our tests is a modem specific to VDSL. The starting point is the implementation of the VDSL modem using discrete components [Mes00]. The schema of the prototype is illustrated in Figure 7.23a. In this schema the ASIC and the FPGA are used for the processing of the fixed signal, the DSP is used for the configurable signal and the MCU for the modem control and the interface with the host PC. The experiments have been done only for the part that is represented in gray in Figure 7.23a. The functionalities of this part are split between two ARM processors and a hardware module in charge with the transmission chain (shown in Figure 7.23b.). This partitioning was suggested by the team that realized the prototype of the VDSL modem [Mes00].

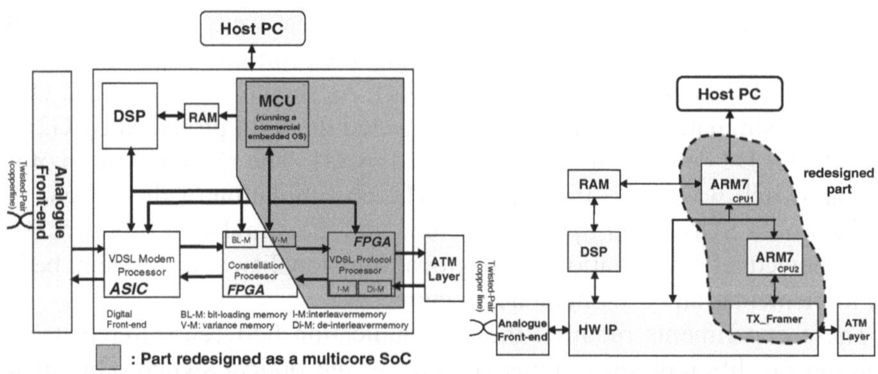

Figure 7.23. VDSL modem – the prototype board (a) and the multiprocessor embedded architecture

6.2 Virtual Architecture

The VDSL modem subset is composed by three modules: M1 and M2 that correspond to the two software modules that in the end are targeted on the ARM processors and M3 corresponding to the hardware module that realizes the transmission chain.

In terms of abstraction levels, the software modules M1 and M2 are implemented at OS level, the module M3 is specified at RTL level and the different modules of the system that communicate through channels, at transaction level. In terms of specification languages, SystemC is used for the specification of different system components.

The virtual architecture equivalent to the VDSL modem is presented in Figure 7.24. As shown, each module of the application is wrapped in a virtual module. The virtual module wrappers VM1 and VM2 connect modules M1 and M2 with the communication channels. In terms of communication protocols, abstraction levels or specification languages, there is no incompatibility between the modules and the corresponding communication channels. The different virtual ports of the wrappers are composed of identical external and internal ports. Meanwhile, these wrappers are necessary in the specification of the abstraction of the hardware/software interfaces that are automatically generated [Lyo01] in order to implement the architecture at RTL level corresponding to the virtual architecture.

The wrapper of the virtual module VM3 connects the module M3 (RTL level) with the communication channels that are at transaction level. This wrapper is composed of internal and external ports that are at different abstraction levels. Each virtual port is composed of internal ports at RTL level and external ports at transaction level. As shown in Figure 7.24, the internal and the external ports are

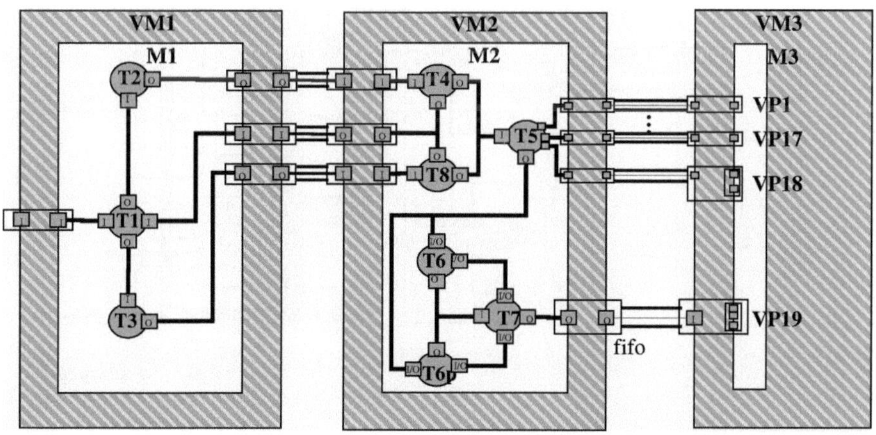

Figure 7.24. Virtual architecture of the VDSL modem

grouped in virtual ports. For instance, the three internal ports of the port VP19 of the virtual module VM3 in Figure 7.24 are specific to the handshake protocol (the request, the data and the confirmation of the end of communication) while the external port is a abstract port that allows the call of communication primitives *Get* and *Put* conform with the protocol FIFO. The port VP19 allows the connection of the module M3, at RTL level, with a channel at a transaction level that embed the protocol FIFO. The same behavior is valid for the other virtual ports of the virtual module VM3. The port VP18 connects the terminal of the locked register protocol to the communication channel at a transaction level and the ports VP1 to PV17 connect a port at RTL level (port that corresponds to the connection through a physical wire) to a channel at transaction level that supplies the communication primitives.

6.3 Results

Multilevel simulation. The execution model for the validation of the presented virtual architecture is illustrated in Figure 7.25. In order to connect the module M3 with the rest of the system, a communication interface needs to be generated. The components from the library necessary for this application are presented in Table 7.4. These elements are characterized by the abstraction level, the protocol, and the direction.

The ports adapters are SystemC modules that are connected with the module through RTL level ports and with the internal communication media through a master port that calls the channel adapter procedures via the internal communication media. After interpreting the RTL level signals that are updated by

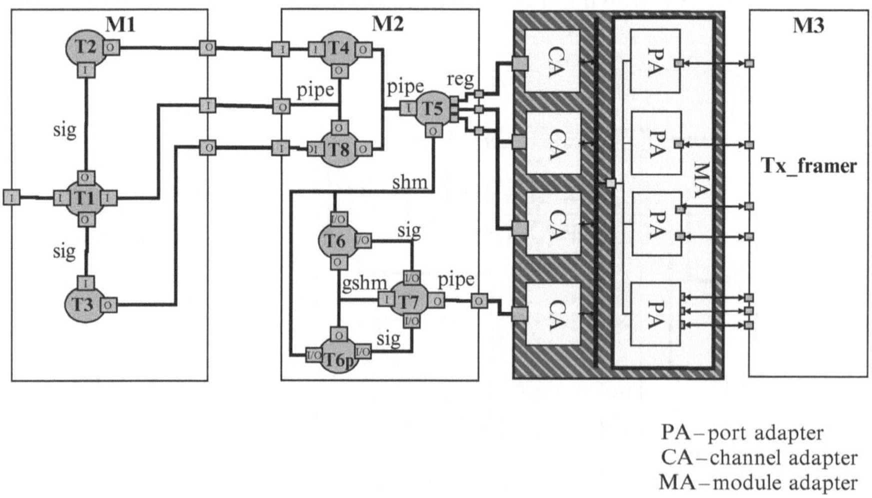

Figure 7.25. Execution multilevel model for the VDSL modem

Table 7.4. Library components used for the multilevel validation of the VDSL modem

Library elements	Abstraction level	Protocol	Direction
Port adapter	RTL	Register	In
			Out
		Guarded register	In
			Out
		Handshake	In
			Out
Channel adapter	Transaction level	FIFO	In
			Out

the module, the port adapter sends the data to the channel adapter via the internal communication media. This action automatically initiates the adaptation function of the channel adapter.

Figure 7.26 shows an excerpt of the SystemC file that was generated. This file specifies the global execution model of the VDSL modem. The figure illustrates only the details concerning a port adapter and a channel adapter, more precisely the port adapter and the channel adapter required to adapt the internal and the external ports of the virtual port VP19. These elements are shown in Figure 7.27. Figure 7.26a illustrates the instantiation of the transaction level channel (ni21_VC21 in Figure 7.27) that has to be connected to three RTL level ports from the module M3 (Req, Data Ack, in Figure 7.27). Figure 7.26b illustrates the instantiation of the submodules of the VDSL modem and the Figure 7.26c the instantiation of the port adapter and channel adapter selected from the co-simulation library. Figure 7.26d shows the adapters interconnections with the module and the channel to adapt. Figure 7.26e corresponds to the code lines necessary to plot the graphs (chronograms). Finally, the code section in Figure 7.26f starts the simulation execution.

The number of code lines as well as the number of interconnections for the initial specification and for the automatic generation of the execution model were evaluated in order to illustrate the advantage of the automatic generation of the simulation models. These numbers are shown in Table 7.5. An important gain (in terms of time and effort) is given by the automatic generation of execution models. The execution model generation time was 90 s.

RTL validation. The RTL level validation architecture for the VDSL modem was also realized, the methodology was used by generating two execution models:

- One execution model that allows the native simulation of the software (the application and the operating system layer). In this case, for each

```
        int sc_main(int ac, char* av[]){
            ch_mac_pipe<long > ni21_VC21;
            sc_signal<bool> Req_VP19;
            sc_ signal <bool> Ack_VP19;
(a)         sc_ signal <bool> Data_VP19;
            sc_mp_link<long int> icm_VP19;
            sc_clocks_clk("s_clk", 20.0, 0.5, 0.0);
            ...
(b)         Ca_in_mac_reg<long int> *Ca_VP19;
            Pa_in_rtl_reg<sc_lv<11> > *Pa_VP19;
            ...
            m1 *M1;
(c)         m2 *M2;
            m3 *M3;
            ...
            M2->VP2_PutVocWord(ni21_VC21);
            Ca_VP19->P(icm_VP19);
            Ca_VP19->Pch(ni21_VC21);
(d)          Pa_VP19->P(icm_VP19);
             Pa_VP19->req(Req_VP19);
             Pa_VP19->ack(Ack_VP19);
             Pa_VP19->data(Data_VP19);
             ...
          sc_trace_file * my_trace_file;
              my_trace_file = sc_create_wif_trace_file("my_trace");
              sc_trace(my_trace_file, s_clk, "clock");
(e)           sc_trace(my_trace_file, Acks_VP19,"Ack");
              sc_trace(my_trace_file, Req_VP19,"Req");
              sc_trace(my_trace_file,Data_VP19,"Data");
              ....
               sc_start(2500);
(f)            return(0);
        }
```

Figure 7.26. Excerpt from the file generated for the multilevel co-simulation

operating system an execution model was used. This execution model for the VDSL modem is illustrated in Figure 7.28a.

- One execution model using instruction set simulators (ISS) for software execution (in this case and in this context, the ARM processor ISS). This

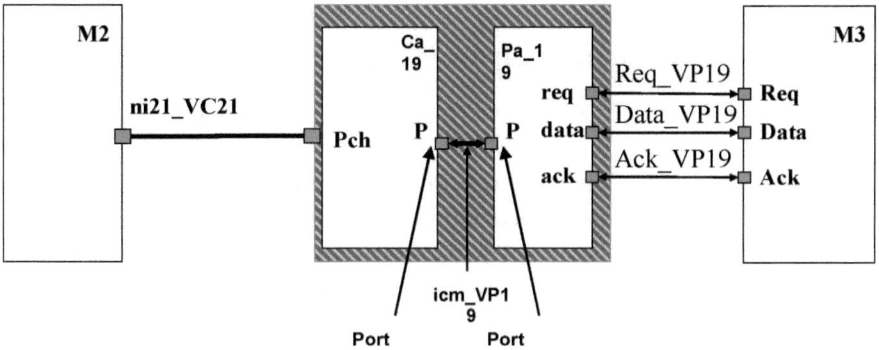

Figure 7.27. The basic interface

Table 7.5. Results for the automatic generation of multilevel execution model for VDSL application

		Code size (number of lines)	Number of interconnections
Virtual architecture specification		150	21
Multilevel	Execution model generation	475	60
	Generation time	90 s	
RTL	Execution model generation	600	187
	Generation time	10 min	

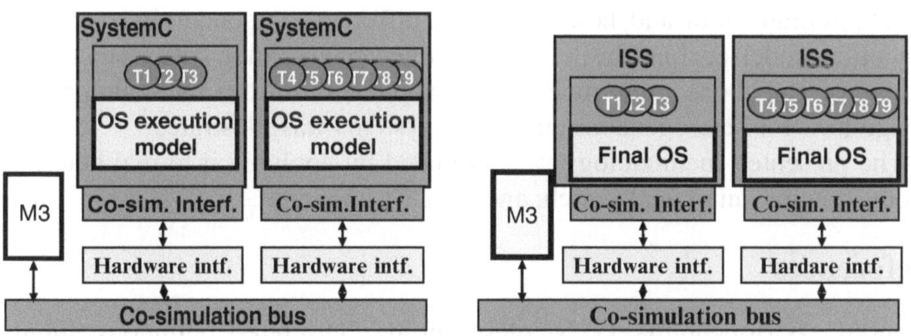

Figure 7.28. Execution models for the validation of the RTL level architecture of the VDSL modem

execution model for the VDSL modem is illustrated in Figure 7.28b. In this case, a BFM is necessary for each processor.

As shown in Figure 7.28, the two execution models have a similar architecture, thus the results in terms of automatic generation of global execution models are comparable. These results are illustrated in Table 7.5. On the other hand, there is a difference in terms of simulation times. The software native simulation produces a gain in speed of two orders of magnitude: the co-simulation with the native software is 100 times faster than the ISS based co-simulation.

6.4 Summary

These experiments allow the validation of the concepts and the methodology that were proposed in this chapter:

- The possibility to use the virtual architecture model for complex applications specification, (applications that are composed by modules described at different abstraction levels)
- The automatic generation of virtual architecture execution models
- The advantage of the simulation model for operating systems – the performance gain in terms of speed

7. Conclusion

This chapter presented a generic simulation model for the heterogeneous embedded systems. In this model, all the components forming the system are connected to the communication network through co-simulation interfaces. The modular structure of these interfaces and the separation between the adaptation of the communication and the adaptation of different components into the global execution model, enabled the definition of a new methodology of automatic generation of global execution models. The application of this methodology in a design flow for heterogeneous embedded systems was illustrated.

The presented methodology was validated by applying it to real heterogeneous systems: an optical switch, and a VDSL modem.

References

[Bou05] F. Bouchhima, G. Nicolescu et al., "Discrete-Continuous Simulation Model for Accurate Validation in Component-Based Heterogeneous SoC Design", 16th IEEE Proc. Int. Workshop on Rapid System Prototyping, June 2005.

[Cad02] Cadance Design Systems, Inc., "Virtual Component Design", available on line at http://www.cadence.com/products/vcc.html

[Cal97] J. P. Calvez, D. Heller, O. Pasquier, "Hardware/Software System Design Based on the MCSE Methodology", *Current Issues in Electronic Modeling*, Vol. 9: *System Design*, Editors J.M. Bergé, O. Levia, J. Rouillard, Kluwer Academic Publishers, 1997, Chap. 6, pp. 115–150.

[Cos99] P. Coste, F. Hessel, Ph. LeMarrec, et al., "Multilanguage Design of Heterogeneous Systems", Proc. Int. Workshop on Hardware-Software Codesign, May 1999.

[Hes00] F. Hessel, P. Coste, G. Nicolescu, Ph. Lemarrec, et al., "Multi-Level Communication Synthesis of Heterogeneous Multilanguage Specification", Proceedings of the IEEE International Conference On Computer Design, September 2000, Austin, USA.

[Kea99] M. Keating, P. Bricaud, *Reuse Methodology Manual*, Kluwer Academic Publishers, 1999.

[Kur01] T. Kurzweg, J. Martinez, S. Levitan, P. Marchand, D. Chairulli, "Dynamic Simulation of Optical MEM Switches", Proc. DTIP France, April 2001.

[Lyo01] D. Lyonnard, S. Yoo, A. Baghdadi, A. A. Jerraya, "Automatic Generation of Application-Specific Architectures for Heterogeneous Multiprocessor System-on-Chip", Proceedings DAC 2001, June 2001, Las Vegas, USA.

[Mat02] A. Tewari, *Modern Control Design with MATLAB and SIMULINK*, Wiley, April 2002.

[Mes00] D. J. G. Mestdagh, M.R. Isaksson, P. Odling, "Zipper VDSL: A Solution for Robust Duplex Communication over Telephone Lines", *IEEE Communication Magazine*, pp. 90–96, May 2000.

[Syn02] Synopsys Eaglei, available at http://www.synopsys.com/products/hwsw/eagle_ds.html.

[Sys02] T. Grotker, S. Liao, G. Martin, S. Swan, *System design with SystemC, Version 2.0*, Springer, May 2002.

[TIA95] TIA/EIA-95A, "Mobile Station-Base Compatibility Standard for Dual-Mode Wideband Spread Spectrum Cellular Systems", 1995.

[Yoo99] S. Yoo, J. Lee, J. Jung, K. Rhe, Y. Cho, K. Choi, "Fast Prototyping of an IS-95 CDMA Cellular Phone : a Case Study", Proc. 6th Conference of Asia Pacific Chip Design Languages, October 1999.

[Wu97] M. C. Wu, "Micro machining for Optical and Optoelelctronic Systems", Proc. IEEE, Vol. 85, No. 11, November 1997.

Index

MIX
Papier aus verantwortungsvollen Quellen
Paper from responsible sources
FSC® C105338

If you have any concerns about our products,
you can contact us on
ProductSafety@springernature.com

In case Publisher is established outside the EU,
the EU authorized representative is:
Springer Nature Customer Service Center GmbH
Europaplatz 3, 69115 Heidelberg, Germany

Printed by Libri Plureos GmbH
in Hamburg, Germany